Symmetry in Chemistry

Symmetry in Chemistry

H. H. Jaffé & Milton Orchin

Professors of Chemistry, University of Cincinnati

John Wiley & Sons, Inc., New York · London · Sydney

Preface

The idea for the present book first took form when we were reviewing and rewriting the material discussed in Chapter 4 of our *Theory and Applications of Ultraviolet Spectroscopy* (John Wiley and Sons, New York, 1962). Although the classical topics involving symmetry considerations, such as optical activity, dipole moments, and conformational analysis, are usually adequately treated in chemistry texts used by students before they begin their first or second year of graduate work, we were struck by the paucity of comprehensive but essentially nonmathematical treatments of symmetry and group theory in such texts. We were convinced that, because symmetry arguments are such a powerful tool in the teaching of such concepts as hybridization, group and molecular orbitals, selection rules in absorption spectroscopy, and crystal structure, a small book devoted exclusively to symmetry in chemistry, developed in an essentially nonmathematical way, could be a great aid to students and research workers interested in these subjects. The growing importance of absorption spectroscopy, the enhanced status of structural inorganic chemistry, and the increasing popularity of molecular-orbital theory all indicate that an understanding of symmetry and elementary group theory will be an essential ingredient of the background of an educated chemist. We are convinced that most chemists will soon use point group and symmetry species notation with the same ease and understanding that practicing chemists now use conformational notation.

With these thoughts in mind, we started to organize the material for this book in November, 1962. Shortly after we began our work, the superb little book by F. A. Cotton[1] appeared. Because his book

[1] F. A. Cotton, *Chemical Applications of Group Theory*, John Wiley and Sons, New York, 1963.

covered many of the topics we had intended to cover, especially the application of group theory to electronic spectra, we shifted our emphasis a bit to make our book more complementary. Nevertheless, we did not depart from our basic objective, which was to provide a comprehensive but nonmathematical treatment of symmetry and its manifestations and applications.

After completing our first three chapters, which are completely nonmathematical, the reader will be able to classify readily almost any molecule under discussion into its appropriate point group. The technique of stereographic projections for symmetry elements in the various point groups is used extensively, because in our experience this technique facilitates the verification of multiple symmetry operations and is also useful for determining the classes of symmetry operations and order of the point group. In Chapter 4 we discuss translations and rotations of molecules under the symmetry operations appropriate to them. This discussion leads into the topics of symmetry species and character tables and the classification of normal vibrations and molecular orbitals. An elementary treatment of vectors and matrix algebra is included for readers who wish this refresher or for those who have not been exposed to these subjects. This discussion is kept to the bare minimum that will provide the background required for a qualitative understanding of group theory. In Chapter 5 we discuss some applications of symmetry and simple group theory to the generation of group orbitals, to the determination of the number of normal vibrations, and to the problem of selection rules in absorption spectroscopy. Chapter 6 is devoted to crystal symmetry, a subject seldom treated in "chemical texts." In all chapters, digressions from the main arguments or discussion of the elementary mathematical concepts are set in small print so that readers familiar with such material can skip these sections without interrupting the main train of thought. The appendices provide complete character tables, tables giving the number of normal vibrations in various symmetry species, and tables showing the direct sums of excited states and combinations states of degenerate vibrations.

The book is intended as a supplementary text for both undergraduate and graduate students who seek a broad background for understanding structural problems. Thus use of the book can be synchronized with the growth of their interest and understanding.

We hope the book will be particularly appealing to the large group of practicing chemists who completed their formal education before the applications of symmetry and group theory became so popular and so necessary for an understanding of the modern literature.

We wish to acknowledge the help, suggestions, and criticisms of our students and colleagues. Lloyd Jones, David Beveridge, and Al Miller merit special thanks, and Professors Joseph Todd, Eugene Corey, Darl McDaniel, and Don Whitman (Case Institute) read some of our chapters and gave helpful suggestions. We also wish to thank Sharon Butrymowicz for expert typing and much help with the illustrations.

Cincinnati, Ohio
January, 1965

H. H. JAFFÉ
MILTON ORCHIN

Contents

Symmetry in Chemistry

1

Introduction

··

1.1 SYMMETRY AS A UNIVERSAL THEME

Nature provides countless examples of symmetry, one of the most pervasive concepts in the universe. Moreover, all the forms of expression that man has evolved utilize the concept in more or less sophisticated constructions. In the nonmathematical sense, symmetry is associated with regularity in form, pleasing proportions, periodicity, or a harmonious arrangement; thus it is frequently associated with a sense of beauty. In the geometric sense, symmetry may be more precisely analyzed. We may have, for example, an axis of symmetry, a center of symmetry, or a plane of symmetry, which define respectively the line, point, or plane about which a figure or body is symmetrical. The presence of these symmetry elements, usually in combinations, is responsible for giving form to many compositions; the reproduction of a motif by application of symmetry operations can produce a pattern that is pleasing to the senses.

Some elementary examples of the patterns that may be developed by the application of symmetry operations can be illustrated in a simple way by using the number 7 as a motif. We may, for example, write down the number 7 and reflect it across a vertical line, obtaining the pattern shown in Fig. 1.1. Each 7 appears to be the mirror image of the other; that is, if one 7 looked into a mirror, the image it would see would appear as shown. The reflection of an object across a symmetry line is one of the possible operations of repetition. If we were to see Fig. 1.1 without the vertical line, we would say that

1

the figure had a plane of symmetry, the plane passing through the line shown in Fig. 1.1 and perpendicular to the plane of the paper. This is the symmetry of left and right, often called bilateral symmetry. It is the symmetry of the left and right hands, the left and right ears, and of many other parts of the human and animal body. If now we invert one of the 7's, as in Fig. 1.2, the resulting arrangement no longer has the plane of symmetry which was present in the arrangement of Fig. 1.1. It has a new symmetry element, an axis of symmetry. This axis may be visualized as being perpendicular to the paper and passing through the point, shown in Fig. 1.2, midway

Fig. 1.1   A plane of symmetry.

Fig. 1.2   An axis of symmetry.

Fig. 1.3   A four-fold axis of symmetry.

Fig. 1.4   Translational symmetry along a line.

between the two 7's. If we now rotate the arrangement of 7's shown in Fig. 1.2 by 180° in either direction around the vertical axis just described, we obtain an arrangement indistinguishable from the arrangement before rotation. This rotation is another one of the possible operations of repetition. In Fig. 1.2 we obtained the repetition by rotation by 180°; in Fig. 1.3 we obtain the repetition by rotation by 90° around a vertical axis passing through the point shown. Rotational symmetry is the symmetry of a regular polygon, of starfish, of flowers, and in general of things growing from a horizontal surface.

Not only can we reflect and rotate our motif, the number 7, in order to repeat it, but we can translate, or move, it in space along a straight line. Thus, for example, we may have a linear array of 7's shown in Fig. 1.4. This is translational symmetry or the symmetry of equidistant segments; it is found in the lower plants and animals, as for example in the segments of a worm or the legs of a centipede. If this linear array of 7's is repeated by translation in the upward direction, we obtain the pattern shown in Fig. 1.5.

Finally, it is possible to add a translation in a third direction, namely, normal to the plane, and this translation repeats the entire plane at equal intervals resulting in the pattern shown in Fig. 1.6.

Now if each of the 7's in Fig. 1.6 is replaced by a point, the resulting collection of points appears as shown in Fig. 1.7. This collection of points is called a space lattice or a three-dimensional lattice array.

```
7 7 7 7 7
7 7 7 7 7
7 7 7 7 7
7 7 7 7 7
```

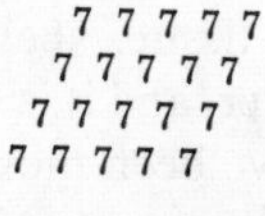

Fig. 1.5 Translation
symmetry in a plane.

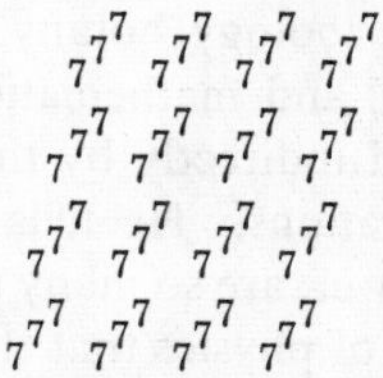

Fig. 1.6 Translational symmetry in three direcions.

The development of this theme would lead us into crystal structure and crystallography; we reserve this development for Chapter 6. What we would like to emphasize at this point is that the repetition of a motif by symmetry operations can lead to more and more intricate patterns, and that these symmetry patterns frequently lie at the heart of much of what we consider pleasing and perhaps even beautiful, whether it be in the crystal habit of a mineral, the cut of a precious stone, the development of a theme in a symphony, or in part of the appeal of a beautiful sculpture. The universal applicability of symmetry considerations is effectively stated by James R. Newman[1] in his introduction to the reprinted lectures on symmetry given by the great mathematician Herman Weyl:

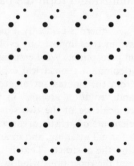

Fig. 1.7 A three-dimen-
sional lattice array.

Symmetry establishes a ridiculous and wonderful cousinship between objects, phenomena and theories outwardly unrelated: terrestrial magnetism, women's veils, polarized light, natural selection, the theory of groups, invariants and transformations, the work habits of bees in the hive, the structure of space, vase designs, quantum physics, scarabs, flower petals, X-ray interference patterns, cell division in sea urchins, equilibrium positions of crystals, Romanesque cathedrals, snowflakes, music, the theory of relativity.

[1] J. R. Newman (ed.), *The World of Mathematics*, Simon and Schuster, New York, 1956, Vol. 1, p. 670.

As expected, both classical and modern literature and poetry make many references to symmetry, both as a universal and Godly theme and as the basis of the beauty of an object.

In his remarkable book[2] D'Arcy W. Thompson synthesizes growth and form in zoology, botany, chemistry, geology, physics, astronomy, engineering, and mathematics in terms of geometry and thus both directly and indirectly by the application of the principles of symmetry operations. He tells us, "Cell and tissue, shell and bone, leaf and flower, are so many portions of matter, and it is in obedience to the laws of physics that their particles have been moved, molded, and conformed. They are no exceptions to the rule that 'God always geometrizes.' "

John R. Platt, who has given considerable thought to the physics of perception and the processes involved in aesthetic evaluation, points out that the kinds of symmetry possible in nature are quite limited and form the basis of perception. He says[3]

... there is evidently a happy coincidence between many of the physiological symmetries imposed by evolution and the primitive pattern symmetries involved in perception. Weyl emphasizes this agreement. Both of these symmetries have a common geometrical explanation. The reason is that they are both derived from the limited set of group-theory symmetries of the "translation groups" and "rotation groups" in three dimensions. We can go on to see that this coincidence of the two sets of forms makes them doubly significant for us as elements of artistic organization, because they now have *both* a referential and a formal meaning, and in art they satisfy us on two levels at once, the biological and the abstract. That is to say, the bizarre fascination of fences and lattices and the repetitious windows of Italian palaces is related on the one hand to the familiar biological repetition of the centipede and the spinal column and on the other hand to the importance of equidistance in our own visual organization of space.

Our discussion of the generality of symmetry in nature and art is intended to be only suggestive and evocative. The books mentioned in this section should be consulted for detailed treatment of the subject.

[2] D'Arcy Wentworth Thompson, *On Growth and Form*, University Press, Cambridge, England, 1915.

[3] J. R. Platt, "Beauty: Pattern and Change," *Functions of Varied Experience*, ed. D. W. Fiske and S. R. Maddi, Dorsey Press, Homewood, Ill., 1961, Chapter 14.

1.2 SYMMETRY IN CHEMISTRY

Our book will attempt to demonstrate the utility of symmetry operations for an insight into various essentially chemical problems. One of the chemical problems always with us is an understanding of structure; how atoms in a molecule are related to each other in space, and how these individual molecules are related to one another in a crystal.

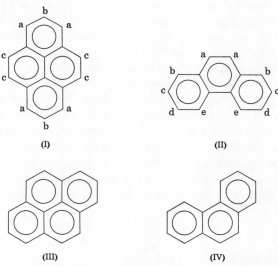

Fig. 1.8   Pyrene (I) and phenanthrene (II) oriented to emphasize the symmetry of the molecules and equivalent positions, and the same molecules, (III) and (IV), oriented according to official rules.

Many advantages accrue to a chemist trained and experienced in the ways of symmetry. One of the least sophisticated consists in the recognition of equivalent atoms in a molecule. Thus the fact that there is only one possible monosubstituted ethane, two possible monosubstituted propanes, and so on, is best taught to under-graduate students on the basis of symmetry considerations. It is often helpful to write the structures of molecules so as to emphasize their symmetry properties. Thus, for example, pyrene and phenan-threne written as in Fig. 1.8, I and II, rather than in the officially approved orientation (Fig. 1.8, III and IV), help bring out the equivalent positions in the molecule. A perspective drawing of a molecule, if drawn well, will usually make its symmetry elements evident.

The advantages to a chemist of a more sophisticated knowledge of symmetry we leave to the reader to assess after he concludes reading this book.

Today many powerful tools can be mustered for the determination of chemical structure. The theory underlying some of these tools relies very heavily on an understanding of symmetry properties. Optical activity, dipole moments, infrared and ultraviolet spectroscopy, and crystal structure all play an essential role in the arsenal available to the chemist; the theory on which each of these is based involves the application of symmetry considerations. Our theme, then, is to show how an understanding of symmetry facilitates an understanding of the theory in each of the fields we mentioned; as a matter of fact, an understanding of symmetry will prove essential. Symmetry arguments are particularly attractive since they can be applied successfully, especially at a qualitative level, without invoking complicated mathematics. What little mathematical theory is required for an understanding of this book, namely, the rudiments of matrix algebra, is developed in the section where matrix transformations are discussed.

1.3   COORDINATE SYSTEMS

In much of our discussion it will be convenient to orient molecules in a Cartesian coordinate system. Practically all conventions for

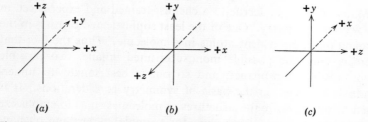

(a)   *(b)*   *(c)*

Fig. 1.9   Three different orientations of the coordinate system, all of which obey the right-hand rule convention.

so doing place the origin of the coordinate system at the center of gravity of the molecule. The designation of the mutually perpendicular x, y, and z axes must be decided, and the conventions for the designation will be developed in the next chapter. In any case, it is then necessary to make a decision as to which side of the zero point of each coordinate should be positive and which should be negative. The convention usually adopted and the one used in this book is the so-called right-hand rule. The thumb, index, and middle fingers of the right hand are extended in three mutually perpendicular directions. The directions in which the thumb, index, and middle fingers are pointing then become respectively the positive x, y, and z directions. Figure 1.9 shows the application of this convention; 1.9a shows a common orientation, 1.9b results from a 90° rotation of the coordinates in (a) around the x axis, and 1.9c results from a 90° rotation of (b) around the y axis. All three coordinate systems are designated so as to obey the right-hand rule.

2

Symmetry Elements
and Symmetry Operations

....

A symmetry operation, such as rotation around a symmetry axis, is an operation that, when performed on an object, results in a new orientation of the object which is indistinguishable from and superimposable on the original. Thus if the molecule HOH is placed in the plane of the paper and rotated 180° about an axis passing through the oxygen atom and bisecting the angle between the hydrogen atoms, as shown in Fig. 2.1, the new orientation is superimposable on the original one; in other words, if a spectator were to close his eyes before the operation was performed, and open them after, it would appear that nothing had been done. In this example the symmetry operation is rotation about an axis; the symmetry element is a rotational axis. The new orientation produced by the 180° rotation is equivalent to but not identical with the original. The two equivalent hydrogen atoms have in fact been interchanged, and the new orientation is indistinguishable from the original because the hydrogen atoms are equivalent and indistinguishable; only if there were some kind of atomic paint for identification would they be distinguishable. A second 180° rotation restores the molecule to the original and identical orientation.

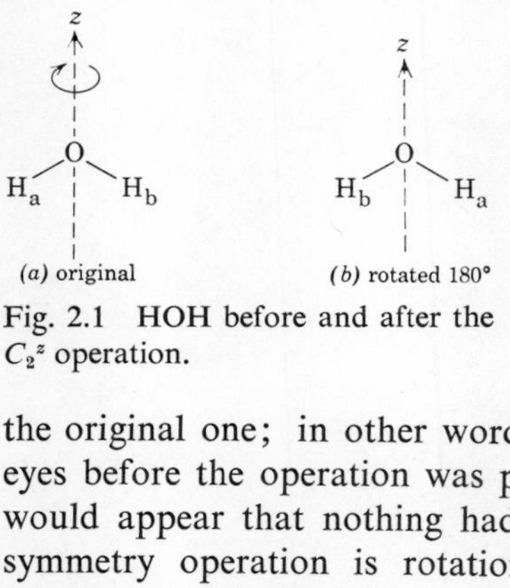

Fig. 2.1 HOH before and after the C_2^z operation.

(a) original (b) rotated 180°

8

The symmetry properties of an object or molecule are best described in terms of the symmetry operations that can be performed on the molecule. Only five basic types of symmetry operations leave the center of gravity unchanged. The five basic operations and the symmetry element associated with each operation will now be described.

2.1 ROTATION ABOUT A SYMMETRY AXIS

If rotation of a molecule about some axis by any angle (the operation) results in an orientation of the molecule that is superimposable on the original, the axis is called a rotational axis (the symmetry element). If the angle through which the molecule must be rotated in order to secure the superimposable image is denoted as $\theta°$, then the molecule is said to have a $360°/\theta$-fold rotational axis, denoted as C_p, where $p = 360°/\theta$ and C stands for cyclic. In the example with the water molecule, a rotation of $180°$ was required to secure the superimposable orientation, and hence the axis of rotation is a $360°/180° = 2$-fold axis, written as C_2. In an alternate system of notation used primarily by crystallographers, the axis is represented by the number of its order, 2 for C_2 axis; the 2 stands for the element.

For many purposes it is desirable to orient a molecule in a Cartesian coordinate system. Although the orientation of a molecule in a coordinate system is inherently arbitrary, certain conventions have evolved which, when adhered to, make communication easier. The rules for selection of the z axis of the molecule are, for the most part, unambiguous, but the assignment of the other axes is frequently arbitrary. Here are the rules for orientation. If followed, they result in an unambiguous assignment of coordinates in practically all cases, and they agree with the practices of most spectroscopists.

1. Place the center of gravity of the molecule at the origin of the coordinate system.

2. For the assignment of the z axis, follow these rules:
    a. If there is only *one* rotational axis in the molecule, this axis is taken as the z axis.
    b. If the molecule has *several* rotational axes, the one of highest order is taken as the z axis.
    c. If there are several rotational axes of the highest order, the axis which passes through the greatest number of atoms is taken as the z axis.

3. For the assignment of the x axis three cases need to be considered; two apply to planar molecules and the third to nonplanar molecules.

  a. If the molecule is planar and the z axis lies in this plane, the x axis is chosen to be the axis *normal* to the plane

  b. If the molecule is planar and the z axis is perpendicular to the plane, the x axis, which must lie in the plane, is chosen so as to pass through the largest number of atoms.

  c. If nonplanar molecules have one plane containing a larger number of atoms than any other plane, they are treated as if they were planar, and as if this preferred plane were the plane of the molecule. If a decision cannot be made on this basis, it will usually be immaterial as to how the assignments of the x and y axes are made. Examples are given in Fig. 2.2.

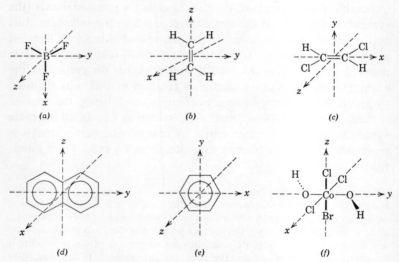

Fig. 2.2   Orientation of various molecules.

Every molecule has an infinite number of C_1 axes. Rotation by 360° around any axis restores the molecule to the original orientation, and hence the C_1 operation around a single-fold rotational axis is a trivial operation around a trivial symmetry element.

Examples of molecules with p-fold rotational axes greater than twofold are shown in Fig. 2.3. In perspective drawings, heavy wedged lines represent bonds projecting out of the plane of the paper, and broken lines represent bonds projecting behind the plane of the paper. Chloroform (Fig. 2.3*a*) has a threefold axis (C_3 or *3*), the axis coincident with the C—H bond. Clockwise rotation of 120° about this axis places Cl^1 at Cl^2, Cl^2 at Cl^3, and Cl^3 at Cl^1, and the resulting orientation is equivalent to, and hence superimposable on, the original one. $PtCl_4^{2-}$ (Fig. 2.3*b*) has a C_4 (*4*) axis perpendicular to the plane of the ion and passing through the Pt atom, while the cyclopentadienide anion (Fig. 2.3*c*) and benzene (Fig. 2.3*d*) have C_5 and C_6 rotational axes, respectively, each perpendicular to the plane of the molecule and passing through its center.

The C_4 axis in $PtCl_4^{2-}$ is also a C_2 axis, and the C_6 axis in benzene is both a C_3 and a C_2 axis. In addition to the C_4 axis and the C_2 axis coincident with it, there are in $PtCl_4^{2-}$, in the plane of the molecule, two C_2 axes which pass through opposite Cl atoms, and two additional C_2 axes, indicated by dotted lines at right angles to each other, and bisecting adjacent Cl-Pt-Cl angles. In $C_5H_5^-$ there are, in addition to the C_5 axis, five C_2 axes in the plane of the molecule; each coincides with one C—H bond and bisects the C—C bond opposite; these five axes are shown by dashed lines.

A linear molecule such as nitrogen (Fig. 2.3*e*), in which all the atoms lie on a straight line, can be rotated around the lengthwise axis passing through the center of all the atoms by any angle imaginable, and hence all such linear molecules have an axis with $p = \infty$ called a C_∞ axis.

2.2 INVERSION (OR REFLECTION) AT A CENTER OF SYMMETRY

If, in a molecule, a straight line drawn from every atom through the center of the molecule and continued in the same direction encounters an equivalent atom equidistant from the center (the operation), the molecule possesses a center of symmetry (the element) designated as *i*, for inversion. No equivalent crystallographic notation is used for *i*, since, as discussed later, crystallographers use a different but equivalent symmetry element. Each atom is thus reflected through the center into an equivalent atom, and therefore atoms must occur in pairs (with the exception of any atom which may lie on the center itself), equidistant but in opposite directions from the center of the molecule, if the molecule is to possess a center of symmetry. Of the molecules shown in Fig. 2.3, only $PtCl_4^{2-}$, benzene, and N_2 have centers of symmetry. If a center of symmetry exists, it lies at the intersection of the three coordinate axes, as in ethylene (Fig. 2.2b). Other molecules possessing a center of symmetry are shown in Fig. 2.4.

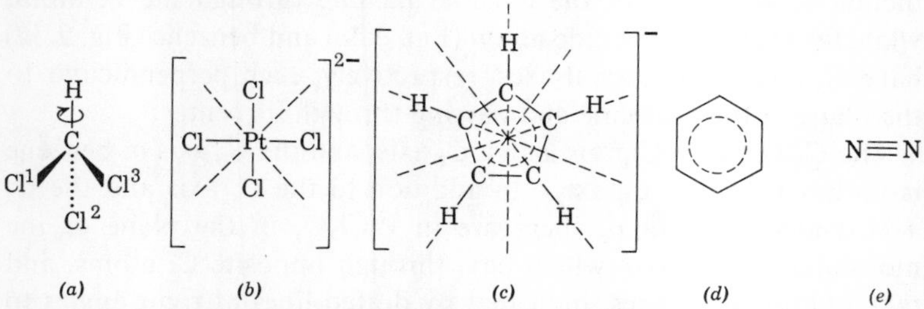

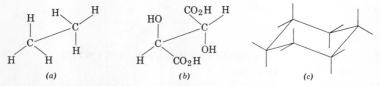

<table>
<tr><td>(a)</td><td>(b)</td><td>(c)</td><td>(d)</td><td>(e)</td></tr>
</table>

Fig. 2.3 Compounds with rotational axes greater than twofold.

Although we have focused our attention on the atoms of a molecule, it is important to realize that an inversion (the operation i) converts every point (x,y,z) into a point $(-x,-y,-z)$. Thus, in the staggered ethane of Fig. 2.4a, placed in the coordinate system as shown in Fig. 2.5, reflection of H^3, which is at a point specified by

<table>
<tr><td>(a)</td><td>(b)</td><td>(c)</td></tr>
</table>

Fig. 2.4 Molecules possessing a center of symmetry. (a) Staggered conformation of ethane; (b) staggered, *trans* conformation of *meso*-tartaric acid; (c) chair form of cyclohexane.

the coordinates x, y, and z, through the center of the molecule gives H^2, which is located at $(-x,-y,-z)$. Similarly, the reflection of H^1 into H^4 and H^5 into H^6 results in the inversion of sign of all three coordinates in each case. Thus, for example, in Fig. 2.5 atoms H^5 and H^6, a pair of atoms exchanged by i, are in the same plane, namely the yz plane, and at first glance it may seem that H^5 and H^6 do not have different signs for the x coordinate, since if $x = 0$, $-x = 0$. If, however, we focus on a point on the periphery of an atom, assuming for our purposes that the atom is a sphere, we see that a point on H^5 toward the *rear* of H^5 has x positive, and upon i this would invert to a negative and equal x, an equivalent point on the *front* of H^6.

At this point it may be profitable to examine the effect of the C_2 operation on the sign of the coordinates. The staggered form of ethane shown in Fig. 2.5 has three C_2 axes, one of which is coincident with the x axis as drawn; the $C_2{}^x$ operation[1] transforms H^1 into H^2, H^3 into H^4, and H^5 into H^6. H^3 may again be represented by (x,y,z); $C_2{}^x$ transforms this point into $(x,-y,-z)$. In general then,

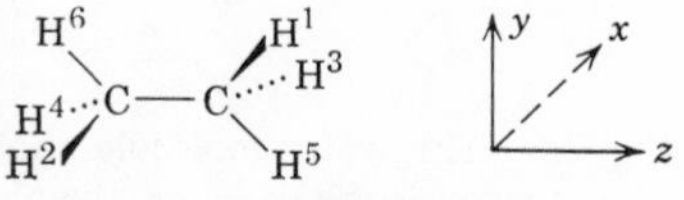

Fig. 2.5   Staggered ethane.

a C_2 rotation around x inverts the sign of every y and z but leaves x unchanged; a $C_2{}^y$ inverts the signs of x and z and leaves y unaffected, and a $C_2{}^z$ leaves z unaffected and inverts x and y. All points on the symmetry element are left unchanged by the symmetry operation.

2.3 REFLECTION AT A PLANE OF SYMMETRY (MIRROR PLANE)

If a molecule is bisected by a plane, and each atom in one-half of the molecule is reflected through the plane and encounters a similar atom in the other half (the operation), the molecule is said to possess a mirror plane (the symmetry element). The mirror plane is designated as σ, or, by crystallographers, m for mirror. For example, in the water molecule (Fig. 2.6), the xz plane (σ^{xz}) is a mirror plane. This mirror plane contains the $C_2{}^z$ axis. There is a second mirror plane in HOH (Fig. 2.6); namely, the plane in the plane of the paper, σ^{yz}. This mirror plane also includes the $C_2{}^z$ axis. Since the z axis is vertical, the two mirror planes, σ^{xz} and σ^{yz}, are *vertical* planes, denoted as σ_v. Note that reflection in the xz plane, for example, converts (x,y,z) to $(x,-y,z)$. The signs of points in the plane do not change by reflection in that plane; only the y changes on the operation σ^{xz}. A molecule may have any number of planes of

[1] The superscript indicates the coordinate coincident with the C_p axis under discussion.

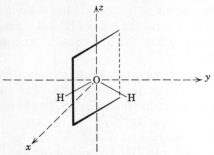

Fig. 2.6   The water molecule showing one of the vertical mirror planes, σ^{xz}. The other σ_v is the yz plane, the plane of the molecule.

symmetry.   A linear molecule such as carbon monoxide, C≡O, has an infinite number of planes of symmetry, all of which include the C_∞ rotational axis, the internuclear axis.

2.4   ROTATION ABOUT AN AXIS, FOLLOWED BY REFLECTION AT A PLANE NORMAL TO THIS AXIS

Assume that a molecule is rotated around an axis and the resulting orientation is reflected in a plane *perpendicular* to this axis (the operation); if the resulting orientation is superimposable on the original, the molecule is said to possess a rotation-reflection axis (the element).   This is the axis around which the rotation was performed, and it is designated as S_p, or by crystallographers as $\tilde{p}$ (read p-tilde). If *trans*-dichloroethylene (Fig. 2.7a) is rotated around the x axis by 180° and the resulting orientation (Fig. 2.7b) is then reflected in the yz plane, the final orientation (Fig. 2.7c) is equivalent to the original,

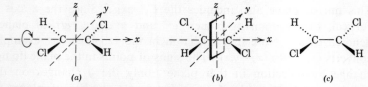

Fig. 2.7   The rotation-reflection operation applied to *trans*-dichloroethylene. (a) The original; (b) after 180° rotation about x; and (c) after reflection of (b) in the yz plane.

and hence the x axis is a rotation-reflection axis. Since the rotation was by 180°, $p = 360°/180° = 2$, and the x axis is an S_2 (or $\tilde{2}$) axis. The rotation-reflection axis is also called an alternating axis because equivalent atoms lie alternately on one side and the other of the plane of reflection.

Every molecule with a plane of symmetry has an S_1 (or $\tilde{1}$) axis, an axis perpendicular to the plane of symmetry. The chemist is usually trained to recognize the plane of symmetry rather than the S_1 axis, especially if this is the only symmetry element present in a molecule.

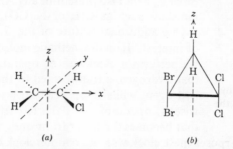

Fig. 2.8   Molecules with a plane of symmetry.

Thus in chloroethylene (Fig. 2.8a) the plane of symmetry is the plane of the molecule, the xy plane, and the z axis is the S_1 axis; rotation of 360° around z and reflection in the perpendicular xy plane gives the original orientation. Of course a rotation of 360° around any axis restores the molecule, but the condition of the S_1 operation requires a rotation followed by a reflection in a plane perpendicular to the rotational axis, and thus x and y are not S_1 axes. 1,1-Dibromo-2,2-dichlorocyclopropane (Fig. 2.8b) has $\sigma = S_1$ as the only symmetry element present; here the z axis is easily recognized as the S_1 axis.

If a molecule has a center of inversion, i, it must also necessarily possess an S_2 axis. We say then that i implies S_2 (and vice versa) because when one of the two is specified, the other is implied and need not be explicitly stated. Actually a center of symmetry, i, implies that any axis through the center is an S_2 ($\tilde{2}$) axis, and a molecule with i has an infinite number of such axes.

Many molecules possess one or more S_p axes, and examples can be found for a variety of values of p. Thus, for example, meth'ane possesses an S_4 axis (actually three of these; Fig. 2.9). Each of these axes passes through the center of the molecule and bisects pairs of opposite HCH angles. Rotation of the molecule by 90° around the vertical axis followed by reflection in a plane perpendicular to this axis restores the molecule to an orientation superimposable on the original. Since there are six equivalent faces to the cube in which the

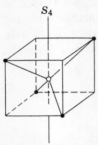

tetrahedral CH_4 may be inscribed (Fig. 2.9), three axes including the one already described, are S_4 axes. Note that each S_4 axis is also a C_2 axis. Hence, in addition to three S_4, there are three C_2 axes also present in any AB_4 tetrahedral molecule such as CH_4 or $Ni(CO)_4$.

One additional feature of the S_4 operation is of interest. If on the methane molecule (Fig. 2.9) we perform a clockwise S_4 operation, the right rear hydrogen is transformed to the lower front hydrogen, while if we perform a counterclockwise S_4 operation, it is the left front hydrogen that becomes the lower front one. Although for

Fig. 2.9   Methane, showing one of the three S_4 axes.

the present purposes these clockwise and counterclockwise S_4 operations give equivalent but not identical orientations, when we discuss certain other kinds of transformations rather than simple atom transformations the distinction becomes important. Accordingly, the

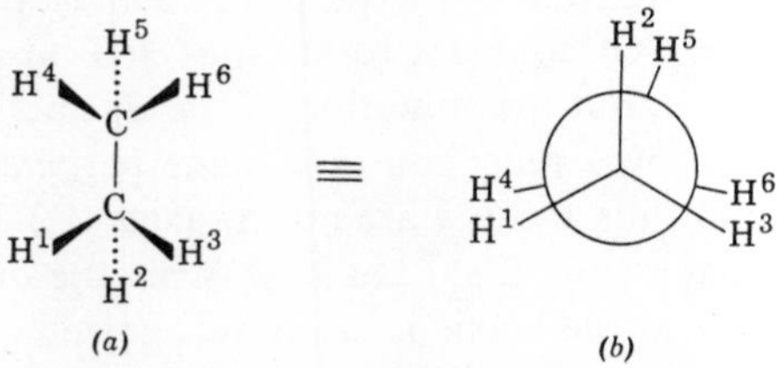

$$H^4 \quad {}^{H^5} \quad H^6 \qquad\qquad {}^{H^2}{}_{H^5}$$
$$C$$
$$\equiv$$
$$C$$
$$H^1 \quad {}_{H^2} \quad H^3 \qquad\qquad {}^{H^4}_{H^1} \qquad {}^{H^6}_{H^3}$$

(a) (b)

Fig. 2.10 (a) A perspective view of eclipsed ethane and (b) an end-on projection view to illustrate that $C_3 \times \sigma_h = S_3$.

clockwise and counterclockwise operations are sometimes denoted as S_4 and S_4' respectively, and represent distinct operations.

There is another interesting property of S_p axes. For S_3 (see the eclipsed ethane of Fig. 2.10), a single clockwise application of S_3 transforms H^1 into H^5, H^2 into H^6, and H^3 into H^4. A second application transforms H^1 via H^5 into H^3, H^2 via H^6 into H^1, and H^3 via H^4 into

H^2; that is, it is equivalent to a counterclockwise application of C_3. A third application of S_3 then transforms H^1 via H^5 and H^3 into H^4, H^2 via H^6 and H^1 into H^5, and H^3 via H^4 and H^2 into H^6, and thus is equivalent to σ_h, a σ perpendicular to the S_3 axis. This implies, then, that S_3 is not an independent operation, but can occur *only* as a combination of C_3 and σ_h. The same is the case for any S_p with p odd, but not with p even.

A great deal of space has been devoted to the S_p operation because of its importance in determining optical activity; this relationship is discussed below.

2.5 ROTATION-INVERSION AXES

The rotation-reflection axis S_p is a complex symmetry element consisting of a series of two operations, neither of which *alone* is necessarily a symmetry operation but which *together* represent a symmetry operation. The same overall operation can be broken down into two other parts, one a rotation, the other an inversion, and this complex operation is called *rotation-inversion*. Although any S_p axis is equivalent to *some* rotation-inversion axis, the order (fold) is not the same. Thus a center of inversion, $S_2 (\tilde{2})$, is equivalent to a onefold rotation-inversion axis, $\bar{1}$ (one-bar) and again *any* axis chosen is a $\bar{1}$ if a center exists. A plane of symmetry, S_1 or $\tilde{1}$, is equivalent to $\bar{2}$. $S_4{}'$ is equivalent to $\bar{4}$, that is, a counterclockwise operation of S_4 is equivalent to a clockwise operation of $\bar{4}$. Similarly, $S_3{}' \equiv \bar{6}$ and $S_6{}' \equiv \bar{3}$; the latter is illustrated in Fig. 2.11.

Both rotation-reflection and rotation-inversion axes are frequently called improper axes as distinguished from rotational axes, which are called proper axes.

2.6 THE IDENTITY

The identity operation is really not an operation at all. The concept of the identity operation, denoted as I, is introduced for mathematical reasons as will be brought out later. The operation I involves no change in the molecule and is thus a pseudo-operation. Obviously every molecule has the symmetry element I, just as every molecule has the symmetry element C_1, which is equivalent to I.[2] The operation I implies doing nothing to the molecule, but the concept is mathematically useful in the development and application of group theory. Thus, for example, two successive applications of C_2 (that

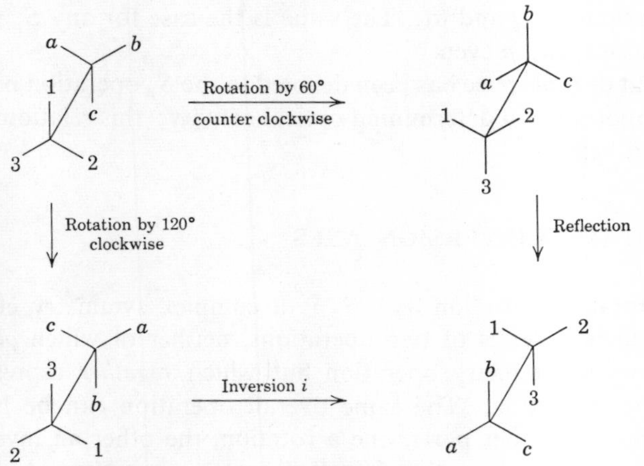

Fig. 2.11 The S_6' and $\bar{3}$ operations, broken down into two steps each, for staggered ethane.

is, C_2^2) restore the molecule or object to its original and *identical* orientation. Hence the operation is equivalent to doing nothing, and we say that $C_2^2 = I$.

2.7 SYMMETRY PROPERTIES OF ORBITALS

We have this far been concerned with symmetry elements and operations as applied to molecules (and all objects) in their stationary states. Symmetry properties are also of importance in connection with the motion of objects or molecules. Although this subject is taken up systematically in Chapter 4, a limited discussion of the symmetry properties of orbitals is of value at this point.

We are familiar with the shapes of orbitals, but for our purposes of the moment we show the angular dependence of some common

[2] This statement is true for all cases considered in this book, but not necessarily in some more complicated cases.

orbital types in Fig. 2.12. Figure 2.12a shows the s orbital, or more correctly, the cut of the spherical orbital in the xz plane. An s orbital is symmetrical with respect to every operation we can think of; it possesses an infinite number of all the symmetry elements we have discussed. The p orbital, shown in Fig. 2.12b, has a node in the yz plane. For our present purposes the significance of the node is that there is a change in sign in going from one lobe of the orbital into the

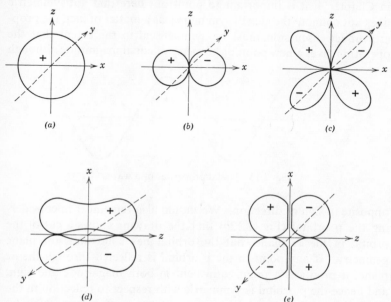

Fig. 2.12   Some common orbital types: (a) s; (b) p_x; (c) d_{xz} atomic orbitals; (d) π; and (e) π^* orbitals in ethylene.

other. This change in sign in passing through the node may be considered as analogous to the change in sign of the amplitude of a wave in passing through a node, as illustrated in Fig. 2.13. For every point in the lobe of the p orbital on the right, which we are considering as minus, there is an equivalent point, equidistant from the center, in the left or plus lobe. Reflection of any point through the yz plane at which the orbital is negative gives an equivalent point at which the orbital is positive.

Whenever a symmetry operation is performed and the point under consideration is transformed to a new point at which the property under discussion has equal magnitude but *opposite* sign, we say the point is antisymmetric with respect to the operation; if the point or property gives a new point of equal magnitude and the *same* sign, we say the point (or property) is symmetric with respect to the operation. In the present case the p orbital shown is antisymmetric with respect to the reflection in the yz plane. Although the relation between symmetric and antisymmetric behavior will be elaborated in Chapter 4, it is important to point out here that antisymmetric does not connote the lack of symmetry; as a matter of fact, the property under discussion must be symmetrical to the extent that the operation gives a new point or property of equal magnitude although

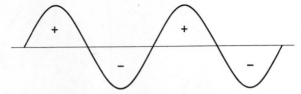

Fig. 2.13   Nodal properties of a wave.

opposite in sign or direction. We should also remember in considering the p orbital (Fig. 2.12b) that the drawing shows a cut of the orbital in the xz plane, but the orbital has solid and not plane geometry. If any point in the p orbital is reflected through the xz plane, it produces a point equivalent in both magnitude and sign, and hence the p orbital is symmetric with respect to reflection in the xz plane as well as symmetric with respect to the xy plane. Now let us consider a C_2 operation around the z axis which is in the nodal plane. Such an operation converts every point in the positive lobe to an equivalent point in the negative lobe and vice versa, and hence this p orbital is antisymmetric with respect to the $C_2{}^z$ operation. The orbital is also antisymmetric with respect to $C_2{}^y$ but symmetric with respect to rotation about the x axis, a C_∞ axis, by any angle. With respect to i, the p orbital is antisymmetric. The d orbital of Fig. 2.12c is symmetric with respect to σ^{xz}, i, and $C_2{}^y$ and antisymmetric with respect to σ^{xy}, σ^{yz}, $C_2{}^z$, $C_2{}^x$, and $C_4{}^y$. Figure 2.12d represents a π orbital as in ethylene. The symmetry properties of this orbital are identical with those of a p$_x$ orbital, except that the nodal plane is the xy plane and the x axis is not now a C_∞ axis.

Finally, Fig. 2.12*e* illustrates an antibonding π orbital (π^* orbital) of ethylene which, as far as symmetry properties are concerned, is very similar to the d orbital shown in Fig. 2.12*c*.

2.8 THE RELATION BETWEEN SYMMETRY AND OPTICAL ACTIVITY

Symmetry becomes a subject of interest to a chemist early in his study of organic chemistry because symmetry (or its absence) is a criterion of optical activity. A discussion of optical activity usually

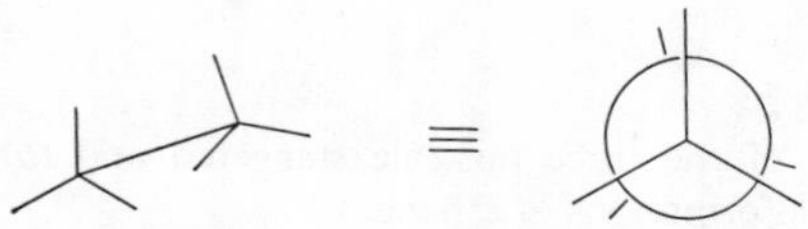

Fig. 2.14   Ethane in the almost-eclipsed, optically active conformation.

immediately precedes the consideration of the carbohydrates. In part, perhaps, this is justified on historical grounds, since carbohydrate chemistry and optical activity were closely intertwined in Emil Fischer's brilliant development of the structure of carbohydrates in the later part of the nineteenth century. Today the use of optically active compounds provides a powerful tool for the elucidation of reaction mechanisms. The relationship between optical activity and the symmetry properties of a molecule thus merits some discussion.

A compound is optically active if its mirror image is not superimposable upon the original. In the application of this test, it must be realized that the atoms in a molecule are in constant motion with respect to each other. Let us consider ethane (Fig. 2.14) in a conformation almost but not quite eclipsed. Now the mirror image of this conformation of ethane is not superimposable on the original one, and hence, by the above test, ethane in this conformation is optically active. If, indeed, all the molecules of ethane were frozen in this conformation, ethane would be optically active. We know, however, that rotation around single bonds, including the C—C bond, is relatively easy and requires only a small activation energy, and hence the conformation shown in Fig. 2.14 is only one of an infinite number of possible conformations. Furthermore, an extremely large number of molecules is present in any practical volume of ethane; for example, approximately 30,000,000,000,000,000 (3×10^{16}) in one cubic millimeter. Consequently for all molecules corresponding to the structure shown in Fig. 2.14, there is a statistically equal probability that a like number of molecules represented by its mirror image are also present. Hence, if the molecule of Fig. 2.14 is arbitrarily considered to rotate the plane of polarized light in a *dextro* direction, the practically equal number of molecules possessing

(a)   *(b)*

Fig. 2.15 (*a*) One of the three possible staggered and (*b*) one of the three possible eclipsed conformations of ethane.

the mirror-image structure will rotate the plane of polarized light an equal amount in the *levo* direction, and hence the net optical activity will be zero; the mixture is a racemic modification.

Although Fig. 2.14 represents a molecule of ethane which is nonsuperimposable on its mirror image, other conformations of ethane, six to be precise, have superimposable mirror images. These conformations are the three possible exactly eclipsed and the three possible exactly staggered conformations; one of each is shown in Fig. 2.15. As will be shown, the fact that (*b*) has a plane of symmetry and (*a*) in addition a center of symmetry immediately implies that the mirror image of (*a*) is superimposable on (*a*) and the mirror image of (*b*) superimposable on (*b*). Consequently we are spared the necessity of writing or drawing the mirror image and performing the super-positioning, although in this particular example the process is simple if not obvious. Ethane can be said to be optically inactive because, for every imaginable conformation, there exists another confor-mation which is the mirror image of the first and because these conformations are readily interconvertible and of equal energy, and accordingly present in equal proportions. Any activity of the first is canceled by that of the mirror image. In only six conformations is the mirror image equivalent and superimposable on the original. *In order to be optically active, a molecule must not be readily able to assume any conformation whose mirror image is superimposable on the original.* Hence it can be immediately concluded that ethane is inactive because it can readily assume a conformation [either (*a*) or (*b*)] which has a superimposable mirror image.

The foregoing discussion indicates how useful it could be to determine whether or not a molecule has a superimposable mirror image, that is, whether it is optically active. If the four groups or atoms attached to the carbon atom in Cabcd are all different, the mirror image of the molecule will not be superimposable on the original, and the carbon atom is called an asymmetric carbon atom. If the molecule is more complicated, it is sometimes difficult to decide whether the four groups attached to a particular carbon are "different." Furthermore, many optically active molecules that contain no "asymmetric" carbon atoms are known. One single criterion is sufficient to determine whether or not a molecule is optically active: the existence of an S_p axis in the molecule. If it has an S_p axis, even if $p = 1$, the molecule is inactive, that is, it is superimposable on its mirror image. In most examples it is common practice to search for a plane of symmetry or a center of symmetry simply because these symmetry elements are easier to spot than the S_1 and S_2 axes to which they correspond, respectively. However, some molecules are known which do not possess either an S_1 or an S_2 axis but which are inactive, owing to an S_p axis of order higher than twofold. Thus, the spiran shown in Fig. 2.16 has neither a plane nor a center of symmetry, but the vertical axis that bisects both rings and goes through the spiro nitrogen atom is an S_4 axis. Rotation by 90° around this axis followed by reflection in a plane perpendicular to the axis and passing through the nitrogen atom restores the molecule to the original configuration. Consequently it is not surprising that it has been impossible to resolve this molecule.

It is important to note that a molecule may possess an element of symmetry and still be optically active. Thus if the only symmetry element present in a molecule is a C_p axis, an ordinary axis of symmetry, the molecule will be optically active. Consider, for example, *trans*-1,2-dichlorocyclopropane (Fig. 2.17). The axis in the plane of the molecule bisecting the methylene group and the opposite C—C bond is a C_2 axis; rotation around it results in an orientation superimposable on the original one. However, the mirror image (*b*) of the molecule is not superimposable on (*a*) and hence (*a*) and (*b*) are enantiomorphs. Note also that the spiran of Fig. 2.16 has a C_2 axis coincident with the S_4 axis; but the presence of the S_4 and not of the C_2 axis is responsible for the lack of optical activity. Molecules such as *trans*-1,2-dichlorocyclopropane, which have a symmetry element but which are optically active, cannot correctly be said to be asymmetric because asymmetric means without symmetry. Such molecules are described as *dissymmetric*. An

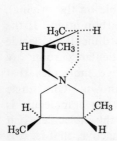

Fig. 2.16 A molecule which is optically in- active because of the presence of an S_4 axis.

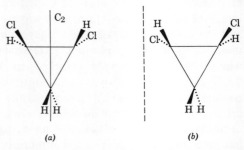

Fig. 2.17 (*a*) An optically active compound and (*b*) its mirror image.

optically active compound need not be asymmetric, but it must be dissymmetric. Thus dissymmetric compounds need not be asym- metric, but all asymmetric compounds are dissymmetric.

It was stated earlier that if any readily accessible conformation of a molecule possesses an S_p axis, the molecule will be inactive. Free rotation around single bonds in most acyclic compounds is possible. The barrier to free rotation would have to be about 15–20 kcal/mole before individual conformers or rotamers could be isolated at room temperature. With cyclic compounds complete rotation around the bonds making up the ring is impossible. However, in the cyclo- hexane and other saturated ring systems a variety of conformations are interconvertible at room temperature. Thus, cyclohexane can exist in either of two (rigid) chair forms which are interconvertible through a (flexible) boat form (Fig. 2.18). In going from one chair form to the other, all the C—C bonds must be partially rotated. The energy barrier for the conversion of a chair to a boat form is estimated to be about 10–11 kcal/mole. The free-energy difference between the more stable chair forms and the boat form is estimated to be about 4 kcal/mole, which is sufficient to insure that about 99.9 percent of the molecules at room temperature are in the chair form. The chair forms of cyclohexane have S_1, S_2, and S_6 axes, the S_6 axis going through the center of the molecule and perpendicular to the approximate plane of the molecule; this axis is also a C_3 axis.

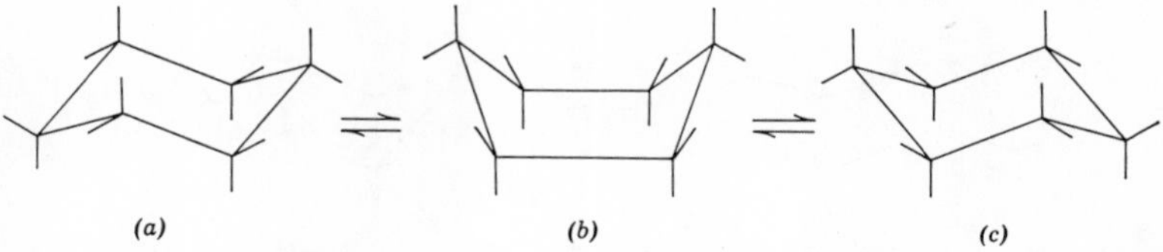

Fig. 2.18   The two interconvertible chair forms (*a*) and (*c*), and the boat form (*b*) of cyclohexane.

If one of the hydrogens is replaced by another group, for example, chlorine (Fig. 2.19), the resulting molecule has an S_1 axis and therefore a plane of symmetry. The plane of symmetry is perpendicular to the ring and includes the chlorine atom and carbons 1 and 4 and the hydrogens attached to these carbon atoms. The S_1 axis is

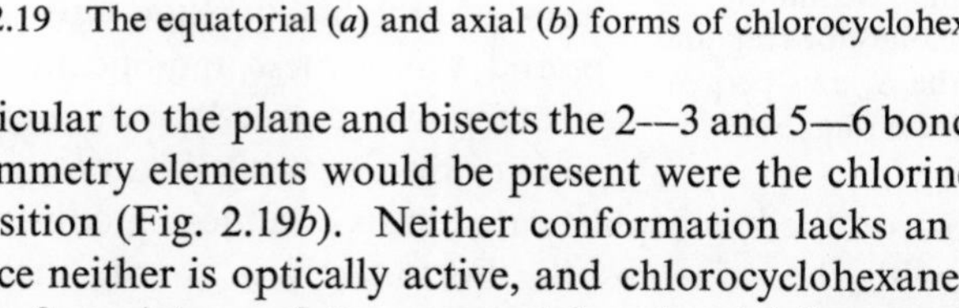

Fig. 2.19   The equatorial (*a*) and axial (*b*) forms of chlorocyclohexane.

perpendicular to the plane and bisects the 2—3 and 5—6 bonds. The same symmetry elements would be present were the chlorine in the axial position (Fig. 2.19*b*). Neither conformation lacks an S_p axis and hence neither is optically active, and chlorocyclohexane, which consists of a mixture of the equatorial and axial forms, although predominantly equatorial, is inactive.

Let us now consider *cis*-1,2-dichlorocyclohexane (Fig. 2.20): In the 1,2-*cis* form, one chlorine is in an axial (*a*) position and the other chlorine is in an equatorial (*e*) position. The inverted form now has the former equatorial chlorine in an axial position, and the former axial in an equatorial. Since both forms are (a, e), they are of equal stability, and since the energy barrier to inversion is not large interconversion occurs readily at room temperature. Thus *cis*-1,2-dichlorethane consists of an equimolar mixture of both forms. Now, neither (a) nor (b) has an S_p axis, and hence we may be tempted to conclude that the compound is optically active. However, close

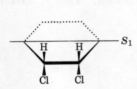

Fig. 2.20 (a) The $1(a),2(e)$-dichlorocyclohexane, and (b) the $1(e), 2(a)$-form.

examination will reveal that Figs. 2.20*a* and 2.20*b* are, in fact, mirror images of each other and thus if (a) were dextrorotatory, any activity would be canceled by the equal population of the equivalently levorotatory enantiomorphic (b). Thus we may consider that *cis*-1,2-dichloro-cyclohexane is inactive because it is a racemic modification. If we were to consider the cyclohexane ring as a completely planar ring (Fig. 2.21), then the possibility of optical activity would be ruled out on the basis of the existence of a plane of symmetry (or S_1 axis); on this basis we might conclude that the compound was a *meso* modification. Thus, whether we correctly rule out optical activity on the basis of a racemic modification or on the less sophisticated basis of a meso modification, we end up with the same result. Fortunately it is true that for the purpose of determining optical activity the cyclohexane ring may be considered planar. Perhaps the theoretical justification for this is that in the ready interconversion of the two chair forms the "average conformation" is a planar ring.

Fig. 2.21   *cis*-1,2-Dichloro-cyclohexane assuming a planar cyclohexane ring and showing the S_1 axis perpendicular to the ring.

Finally, mention should be made of a special situation in which a molecule may possess no S_p axis in any conformation and still be inactive. Such a molecule is (*dextro*)-menthyl-(*levo*)-menthyl-2,6,2′,-6′-tetranitro-4,4′-diphenate (Fig. 2.22). The mirror image of this molecule is not superimposable on the original. However, if the 4,4′ bonds in the mirror image are rotated, then the new molecule is

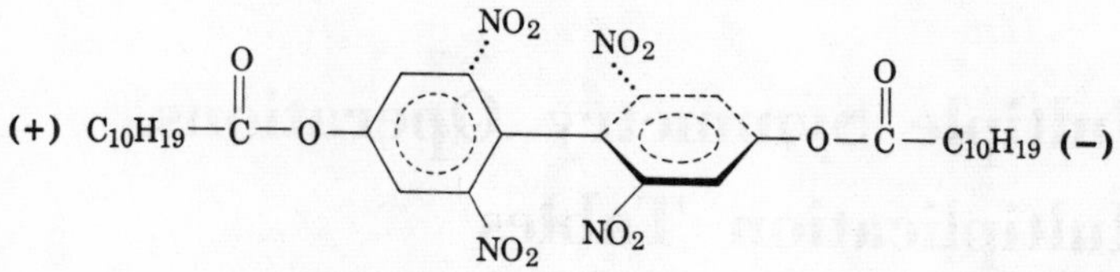

Fig. 2.22   A dissymmetric compound that is inactive.

superimposable on the original, because of the generation of a pseudo S_4 axis. Such rotations are possible under ordinary conditions, and these rotameric interconversions ensure statistically equal populations of enantiomeric conformations and thus account for the observed optical inactivity.

REFERENCES

Robert P. Bauman, *Absorption Spectroscopy*, John Wiley and Sons, New York, 1962, Chapter 10.

Andrew Streitwieser, Jr., *Molecular Orbital Theory for Organic Chemists*, John Wiley and Sons, New York, 1961, Sections 3.5–3.6.

Gerhard Herzberg, *Molecular Spectra and Molecular Structure*, Vol. 2, *Infrared and Raman Spectra of Polyatomic Molecules*, D. Van Nostrand Co., Princeton, N.J. 1945, pp. 1–12.

P. J. Wheatley, *The Determination of Molecular Structure*, Oxford University Press, London, 1959, pages 1–19.

Arnold Weissberger, ed., *Technique of Organic Chemistry*, Vol. 9, *Chemical Applications of Spectroscopy*, Interscience Publishers, New York, 1956, pp. 207–211, 629–631.

Multiple Symmetry Operations.
Multiplication Tables
and Point Groups

In our earlier discussion we pointed out that if a molecule has a C_4 axis, as, for example, square planar $PtCl_4^{2-}$, a C_2 axis is necessarily implied; and an S_4 axis, as in CH_4, also implies a C_2 axis. Similarly, we have shown that an S_3 axis implies *both* a C_3 axis and a σ_h plane.[1] Not only do certain individual symmetry elements suggest other individual elements, but combinations of symmetry elements are equivalent to another single element or combination of elements. Thus in *trans*-dichloroethylene the combination of $\sigma_h \times C_2^z$ is equivalent to i, and this is again readily verified by the fact that C_2^z converts the point (x,y,z) into $(-x,-y,z)$ and σ^{xy} converts $(-x,-y,z)$ into $(-x,-y,-z)$; hence $\sigma_h \times C_2^z = i$. Or, with H—O—H (Fig. 2.6), reflection in the xz symmetry plane transforms y into $-y$; reflection in the yz symmetry plane transforms x into $-x$. The transformation of (x,y,z) into $(-x,-y,z)$ is also achieved by C_2^z; hence $\sigma^{yz} \times \sigma^{xz} = C_2^z$, and thus the existence of two symmetry planes at right angles implies a C_2 axis, the intersection of the two planes.

---

[1] In σ_h the subscript h indicates that the plane is horizontal, and therefore perpendicular to the principal axis, which by convention is always taken to be vertical.

3.1   GROUPS

If we again consider *trans*-dichloroethylene as a model, we see that this molecule has four symmetry elements, that is, four different symmetry operations, each of which, if applied to the molecule,

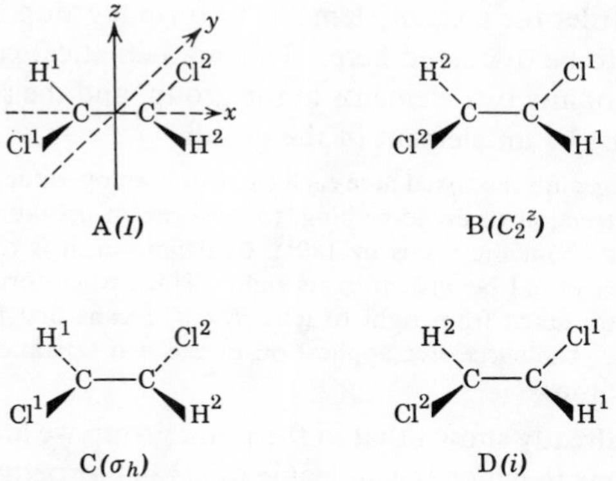

Fig. 3.1   The four configurations associated with the four symmetry operations in the set making up point group C_{2h}.

restores the molecule to an orientation identical or equivalent to the original. This set of four symmetry elements is I, C_2^z, σ^{xy}, and i; the symmetry operations that correspond to these symmetry elements produce the configurations which we have shown in Fig. 3.1.

Configurations B, C, and D are equivalent and superimposable on the original, and of course I is identical with the original. The set of four symmetry elements (or the four symmetry operations) is said to form a group. In all these operations one point, the center of gravity of the molecule, is left unchanged, and hence the group is called a *point group*. Crystallographers are interested in groups involving a translation of the center of gravity; these groups may be line, plane, or space groups. Our discussion at this time will be limited to point groups; the other kinds of groups will be discussed in Chapter 6.

3.2 MULTIPLICATION OF OPERATIONS

Although we will restrict our discussion of groups to sets of symmetry elements forming the group, it must be realized that the group is really a mathematical group. Although we will not go into the properties of mathematical groups, one of the rules that must be satisfied in order for a set of elements to form a group is of sufficient importance to be discussed here. This mathematical rule states that the product of any two elements in the group, and the square of each element, must be an element in the group.

Now, in the groups discussed here each element is an operator, not a quantity; that is, an instruction to do something, to perform an operation; specifically, $C_2{}^z$ says rotate about the z axis by $180°$; by definition, it is to operate on an operand, which would be written to its right. Thus, when forming a product, the elements are taken from right to left: $i \times \sigma_v$ means first to reflect on σ_v, then to invert. The successive application of the two operations is called the product of the two.

We have already shown that in the point group we are considering, namely the one to which the molecule *trans*-dichloroethylene belongs, $C_2{}^z \times \sigma^{xy} = i$. Also then, $i \times \sigma^{xy} = C_2{}^z$ and $C_2{}^z \times i = \sigma^{xy}$, as can be verified most easily by using the change in signs of coordinates technique previously illustrated. In the group considered here, the order of the multiplication does not make any difference; that is, using the letters in Fig. 3.1 for the operations, $A \times B = C$, and $B \times A = C$. In algebra, when $AB = BA$ we say the multiplication is *commutative*.

Although, in the example we chose, the multiplication is commutative, this is not generally true for groups. (Groups for which it is true, like the group under discussion, are called *Abelian*.)

A simple example in which the order of multiplication makes a difference in the result follows (Fig. 3.2). Assume the BF_3 molecule (which is planar) to be oriented with the three fluorines labeled 1, 2, and 3 in the plane of the paper. This molecule has a C_3 axis perpendicular to the plane of the molecule and passing through the center of the molecule, and three σ_v's, each of which includes one of the three BF bonds.[2] It should be appreciated that a C_3 clockwise operation is

[2] In σ_v the subscript v indicates that the plane is vertical; that is, it includes the principal axis.

not identical with a C_3 counterclockwise rotation (denoted as C_3'). However, two successive C_3 operations (that is, $C_3{}^2$) produce a configuration identical with C_3'. Now if we rotate the molecule in orientation (*a*) counterclockwise by 120°, the C_3' operation, we obtain orientation (*b*), and if we then reflect this orientation through the σ_v'' plane, we obtain the orientation (*c*). When we perform the C_3' operation we do not rotate the planes of symmetry; σ_v'' refers

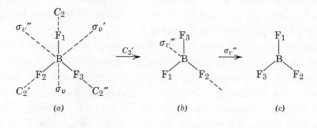

Fig. 3.2 Noncommutative operations showing $\sigma_v'' \times C_3' = \sigma_v \neq \sigma_v' = C_3' \times \sigma_v''$.

to the plane in the same position in both the original and the rotated molecule. Now orientation (*c*) is identical to that obtained by reflecting the original orientation (*a*) through the σ_v plane. In accordance with the notation for the product of symmetry operations, the above operations are indicated as follows: $\sigma_v'' \times C_3' = \sigma_v$. If we reverse the operations, and, starting with orientation (*a*), first perform a σ_v'' operation, we obtain orientation (*b'*); if now we perform a counterclockwise 120° rotation around the C_3 axis, we obtain orientation (*c'*), clearly different from orientation (*c*), but identical to the result of performing σ_v' on orientation (*a*). Accordingly,

$$\sigma_v'' \times C_3' = \sigma_v \neq \sigma_v' = C_3' \times \sigma_v''$$

and the operations do not commute.

3.3   STEREOGRAPHIC PROJECTIONS

The multiplication of symmetry operations can be very effectively deduced and verified by means of a technique called *stereographic projection*. Such projections are very helpful also in many other problems, such as the determination of crystal angles and the deduction of the number of atoms which are symmetrically equivalent, that is, belong to a set, in a given point group. The latter problem is discussed in some detail in Chapter 6.

Although the development of a stereographic projection can (and will) be illustrated by the atom-by-atom projection of a molecule, it is instructive to discuss first the projection of various *symmetry elements* of a molecule. Imagine a molecule placed in a hollow sphere with its center of gravity at the center of the sphere. Then all planes and axes of symmetry have a common intersection at the center. Now the purpose of the stereographic projection is to show all the symmetry elements in a *two*-dimensional drawing, and in order to do so, their intersections with the sphere are projected by a standard procedure onto the equatorial plane of the sphere. The intersection with the sphere of a horizontal plane coincident with the equatorial plane is the circle with the circumference of the equatorial plane. If such a horizontal symmetry plane is present in the molecule, it is indicated in the stereographic projection by drawing the circle as a full line; otherwise, the circle is drawn dashed. A vertical symmetry axis intersects at the north and south poles; these points are projected onto the center of the equatorial plane and are indicated by a small polygon, the number of sides of which indicate the order of the axis. The polygon is drawn in outline if the axis is an alternating (rotation-reflection) axis, otherwise full; thus, an S_4 is drawn as $\square$, a C_4 as $\blacksquare$. A twofold axis is drawn as a two-sided polygon ($\blacklozenge$).

A vertical symmetry plane, that is, a plane at right angles to the equatorial plane, on projection onto the equatorial plane becomes a line and is indicated as a solid line. A horizontal symmetry axis, already lying in the equatorial plane, intersects the sphere on two diagonally opposite points of the circle, and, of course, no projection is involved. The two intersection points are indicated by the appropriate polygon, connected by a dashed line; if the symmetry axis lies in a vertical mirror plane, the connecting line is drawn solid to represent the plane. The symbols for various symmetry elements are illustrated in Fig. 3.3. A center of symmetry, of course, lies in the equatorial plane, but it cannot readily be projected stereographically. Consequently, it is normally indicated by showing a vertical twofold rotation-reflection axis, to which it is equivalent.

All the elements so far shown were either vertical or horizontal and were easy to project. More difficult are elements which have an arbitrary orientation. A slanted axis, for example, is shown by

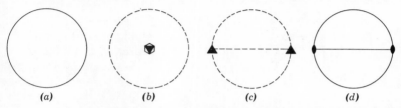

(a)   (b)   (c)   (d)

Fig. 3.3   Symmetry elements in stereographic projection: (a) a horizontal plane: (b) a vertical sixfold rotation-reflection axis and the implied threefold rotational axis; (c) a threefold horizontal rotational axis; and (d) a twofold horizontal rotational axis, a horizontal and a vertical plane.

placing the appropriate polygons at the intersections of the axis with the sphere, projected by the rules to be discussed presently, and then connected by a dashed line. A slanted plane is similarly projected, and comes out as an oval solid line in the projection. This matter will be discussed in more detail on page 50.

The actual construction of a stereographic projection requires first the production of a *spherical projection*. This is achieved by imagining an intense light source at the center of the sphere, and considering the circumscribing sphere as a screen on which the shadow of any object (or point) is observed. The special simplicity of handling the symmetry elements, which made the treatment in the preceding paragraphs desirable, arises because the intersections of the elements with the sphere already represent the spherical projections of the elements.

To further illustrate the construction of the spherical, and hence the stereographic projection, take the actual molecule H_2O_2. This molecule has the geometry of a flat arch with the sides and top intersecting at an angle of $97°$ and,

where the two legs are moved out of the same plane and separated by a dihedral angle of $94°$. The two oxygen atoms are sitting on the ends of a horizontal line in the plane of the paper, and the two hydrogen atoms at the ends of the legs, one in front and the other behind the plane. Now the molecule is placed in a sphere so that the center of gravity of the molecule coincides with the center of the sphere, and an intense light source is placed at this center. Then each atom will produce a shadow on the surface of the sphere. This is the spherical projection and is shown in Fig. 3.4a.

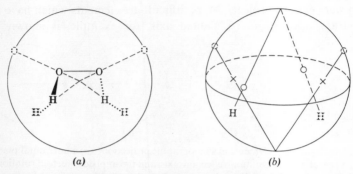

(a)	*(b)*

Fig. 3.4   (*a*) Spherical projection of HOOH; (*b*) projection of the atoms (points) of HOOH in Fig. 3.4a onto an equatorial plane.

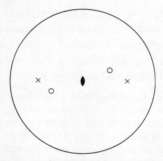

Fig. 3.5   The projection plane showing the position of the intersection of the lines and the plane in Fig. 3.4b.

The spherical projection can be transformed into a planar two-dimensional one as follows. First place an equatorial plane through the sphere; in a drawing on a plane surface this is represented by a horizontal diameter. Now draw a line from the south pole of the sphere to each of the points (oxygen atoms) in the northern hemisphere (Fig. 3.4b). In our example these lines will lie in the plane of the paper since the oxygen atoms lie in this plane. Similarly, connect the north pole with each point (hydrogen atom) in the southern hemisphere; the line from the north pole on the left will start in the plane of the paper and come out of the plane, and the one on the right will go behind the plane, as illustrated in Fig. 3.4*b*. Now we will use the equatorial plane as the projection plane. We place a cross (×) where the lines from the south pole intersect this plane; we also place a small circle (○) where the lines from the north pole intersect the plane. The projection plane that was perpendicular to the paper in Fig. 3.4*a* is now turned 90° and placed in the plane of the paper: it will then appear as in Fig. 3.5, which is a unique projection of any point or atom in the molecule. This projection of points in a sphere onto a plane is the *stereographic projection*; points (atoms) which lie above the equatorial plane appear as crosses, points which lie below as circles.

Inspection of Fig. 3.5 shows immediately the presence of a C_2 axis. It is the axis perpendicular to the plane of the paper and passing through the center of the circle; in the sphere (Fig. 3.4*a*) it is the north pole-south pole axis. The fact that our molecule possesses a C_2 axis might not be nearly as apparent from a consideration of the three-dimensional geometry of the molecule.

In a molecule with many atoms the stereographic projection as drawn here becomes very complicated. Hence a simplification in which the individual atoms are not considered is necessary. The symmetry elements present in the molecule are determined, and the stereographic projection of these elements is drawn as explained above. The symmetry operations are then performed on a general point and the circles and crosses drawn where appropriate, as will be shown below.

The essential information about the symmetry of a molecule, or more generally the point group to which it belongs, can readily be obtained from the stereographic projection. This is done by considering any point in the stereographic projection, and the more general the point, the better. Therefore, the point should not lie on any of the symmetry elements. Such a general point need not be an atom—frequently it is not—it may be any point defined with reference to the symmetry elements of a molecule. This process of drawing stereographic projections is particularly useful in finding the relations between symmetry operations. Take the example of two planes at right angles, illus-

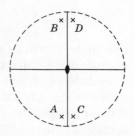

Fig. 3.6 The four equivalent points in an object with two mirror planes.

trated in Fig. 3.6. The general point A of Fig. 3.6 is reflected into B by one plane (σ_1) and into C by the other (σ_2), so that A, B, and C are equivalent points. But C is converted into D by σ_1 and B into D by σ_2, so that a general point must have the four equivalent positions A, B, C, and D. Examination of the figure, however, shows that rotation about a vertical axis moves A into D,

B into C, and vice versa; hence the two planes imply the existence of a C_2 axis. Further, it is readily seen that successive application of σ_1 and σ_2, $\sigma_2 \times \sigma_1$, transferring A to B to D, is equivalent to C_2: $\sigma_2 \times \sigma_1 = C_2$. The use of stereographic projections constitutes one of the most convenient ways of determining the product of symmetry operations, and we return to this technique again shortly.

3.4 MULTIPLICATION TABLES

Every molecule may be characterized by the symmetry elements it possesses or by the symmetry operations we can perform on it. If we list all the symmetry operations applicable to any one molecule, these

Table 3.1 Multiplication Table for the Group in Fig. 3.1

	I	$C_2{}^z$	$\sigma_h{}^{xy}$	i
I	I	$C_2{}^z$	$\sigma_h{}^{xy}$	i
$C_2{}^z$	$C_2{}^z$	I	i	$\sigma_h{}^{xy}$
$\sigma_h{}^{xy}$	$\sigma_h{}^{xy}$	i	I	$C_2{}^z$
i	i	$\sigma_h{}^{xy}$	$C_2{}^z$	I

operations constitute what the mathematician calls a group. An example of a group is the four symmetry operations applicable to the *trans*-dichloroethylene molecule; their effect on the molecule was shown in Fig. 3.1. One important property of mathematical groups is that every possible product of two operations in the set is also an operation in the set.[3] We use this property in constructing what are called multiplication tables, and for initial illustrative purposes let us return to the group of Fig. 3.1. We place each of the operations in the first row and again in the first column of Table 3.1. Now, if we multiply an element in the row, say $C_2{}^z$, by an element in the column, say $\sigma_h{}^{xy}$, implying operation $\sigma_h{}^{xy}$ followed by $C_2{}^z$, we will always get another element in the set to which the product is equivalent; in the present example, $C_2{}^z \times \sigma_h{}^{xy} = i$.

[3] For an adequate definition of a group, the reader is referred to textbooks on group theory.

We verified early in this chapter that the product of any two of the elements of this group is a third element in the set by the use of change of sign of coordinates. The multiplication of operations is also readily carried out by employing the stereographic projection technique. Thus a molecule with a C_2 axis and a plane of symmetry. would be represented by a solid circle (σ_h) and a two-angled solid polygon (C_2) at the center (Fig. 3.7c). To begin with, a general point in the sphere, say a point in the northern hemisphere, is indicated by

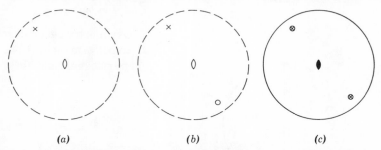

| (a) | (b) | (c) |

Fig. 3.7   Stereographic projection of equivalent points in a molecule possessing i, σ_h, and C_2 (a) the general point; (b) after i; (c) after σ_h showing the four equivalent points and that $\sigma_h \times i = C_2$.

a cross on the stereographic projection (Fig. 3.7a). As a result of the symmetry element i, we know a point equivalent to the first point exists in the southern hemisphere an equal distance from the center and its projection is represented by a circle, as shown in Fig. 3.7b. Finally, since the element σ_h is present, the point both before and after inversion can be reflected in the σ_h plane; hence the cross is circled and the circle is crossed and we get the stereographic projection shown in Fig. 3.7c, indicating that four equivalent points are required by the set of symmetry elements characteristic of the group. Now using the stereographic projection we can again verify that the multiplication of any two elements gives a third. For example, a point represented by the upper cross on inversion gives a point represented by the lower circle; on reflection through σ_h this is converted to the lower cross. This is exactly equivalent to performing a C_2 on the original cross, and henc$e$ $\sigma_h \times i = C_2$. The conventions used for indicating a few different kinds of symmetry operations and the number of equivalent points for such symmetry operations are shown in Fig. 3.8.

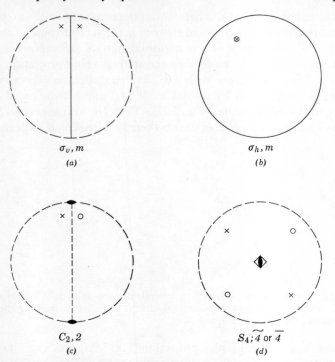

σ_v, m σ_h, m
(a)   (b)

$C_2, 2$ $S_4; \widetilde{4}$ or $\overline{4}$
(c)   (d)

Fig. 3.8 The stereographic projection for some symmetry elements and the equivalent points required by such elements.

The water molecule (Fig. 2.6) has the symmetry elements C_2^z, σ_v^{xz}, and σ_v^{yz}. The multiplication table for this group is shown in Table 3.2; the stereographic projection has already been shown in Fig. 3.6; our earlier discussion showed how the stereographic projection was used to verify $\sigma_1 \times \sigma_2 = C_2$.

Table 3.2 Multiplication Table for Point Group C_{2v}

	I	C_2^z	σ_v^{xz}	σ_v^{yz}
I	I	C_2^z	σ_v^{xz}	σ_v^{yz}
C_2^z	C_2^z	I	σ_v^{yz}	σ_v^{xz}
σ_v^{xz}	σ_v^{xz}	σ_v^{yz}	I	C_2^z
σ_v^{yz}	σ_v^{yz}	σ_v^{xz}	C_2^z	I

As a final example of a multiplication table, we can turn to a more complicated case and consider a molecule like the square planar $PtCl_4^{2-}$. The symmetry operations that can be performed on this ion are shown in Fig. 3.9, and the 16×16 multiplication table constructed from these operations, Table 3.3, defines a group of order 16, the point group D_{4h}.

In molecules with C_p where $p > 2$, the clockwise and counter-clockwise rotations that produce identical configurations are of

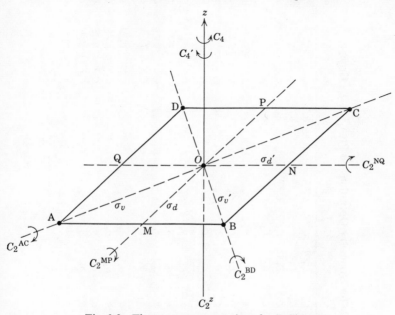

Fig. 3.9   The symmetry operations for $PtCl_4^{2-}$

C_4 = rotation of $+2\pi/4$ about the Oz axis; $S_4 = C_4 \times \sigma_h$

C_4' = rotation of $-2\pi/4$ about the Oz axis; $S_4' = C_4' \times \sigma_h$

$C_2^z = C_4^2 = C_4'^2$; $C_2^z\sigma_h = S_2 = i$

C_2^{AC} = rotation of π about the AC axis

C_2^{BD} = rotation of π about the BD axis

C_2^{MP} = rotation of π about the MP axis

C_2^{NQ} = rotation of π about the NQ axis

σ_h = reflection in the plane (ABCD)

σ_v = reflection in the plane (AC,Oz)

σ_v' = reflection in the plane (BD,Oz)

σ_d = reflection in the plane (MP,Oz)

σ_d' = reflection in the plane (NQ,Oz)

Table 3.3   The Multiplication Table for Symmetry Operations on $PtCl_4^{2-}$ (D_{4h} Group)

D_{4h}	I	C_4	C_4'	C_2^z	C_2^{AC}	C_2^{BD}	C_2^{MP}	C_2^{NQ}	σ_v	σ_v'	σ_d	σ_d'	σ_h	S_4	S_4'	i
I	I	C_4	C_4'	C_2^z	C_2^{AC}	C_2^{BD}	C_2^{MP}	C_2^{NQ}	σ_v	σ_v'	σ_d	σ_d'	σ_h	S_4	S_4'	i
C_4	C_4	C_2^z	I	C_4'	C_2^{MP}	C_2^{NQ}	C_2^{BD}	C_2^{AC}	σ_d	σ_d'	σ_v'	σ_v	S_4	i	σ_h	S_4'
C_4'	C_4'	I	C_2^z	C_4	C_2^{NQ}	C_2^{MP}	C_2^{AC}	C_2^{BD}	σ_d'	σ_d	σ_v	σ_v'	S_4'	σ_h	i	S_4
C_2^z	C_2^z	C_4'	C_4	I	C_2^{BD}	C_2^{AC}	C_2^{NQ}	C_2^{MP}	σ_v'	σ_v	σ_d'	σ_d	i	S_4'	S_4	σ_h
C_2^{AC}	C_2^{AC}	C_2^{NQ}	C_2^{MP}	C_2^{BD}	I	C_2^z	C_4'	C_4	σ_h	i	S_4'	S_4	σ_v	σ_d'	σ_d	σ_v'
C_2^{BD}	C_2^{BD}	C_2^{MP}	C_2^{NQ}	C_2^{AC}	C_2^z	I	C_4	C_4'	i	σ_h	S_4	S_4'	σ_v'	σ_d	σ_d'	σ_v
C_2^{MP}	C_2^{MP}	C_2^{AC}	C_2^{BD}	C_2^{NQ}	C_4	C_4'	I	C_2^z	S_4	S_4'	σ_h	i	σ_d	σ_v	σ_v'	σ_d'
C_2^{NQ}	C_2^{NQ}	C_2^{BD}	C_2^{AC}	C_2^{MP}	C_4'	C_4	C_2^z	I	S_4'	S_4	i	σ_h	σ_d'	σ_v'	σ_v	σ_d
σ_v	σ_v	σ_d'	σ_d	σ_v'	σ_h	i	S_4'	S_4	I	C_2^z	C_4'	C_4	C_2^{AC}	C_2^{NQ}	C_2^{MP}	C_2^{BD}
σ_v'	σ_v'	σ_d	σ_d'	σ_v	i	σ_h	S_4	S_4'	C_2^z	I	C_4	C_4'	C_2^{BD}	C_2^{MP}	C_2^{NQ}	C_2^{AC}
σ_d	σ_d	σ_v	σ_v'	σ_d'	S_4	S_4'	σ_h	i	C_4	C_4'	I	C_2^z	C_2^{MP}	C_2^{AC}	C_2^{BD}	C_2^{NQ}
σ_d'	σ_d'	σ_v'	σ_v	σ_d	S_4'	S_4	i	σ_h	C_4'	C_4	C_2^z	I	C_2^{NQ}	C_2^{BD}	C_2^{AC}	C_2^{MP}
σ_h	σ_h	S_4	S_4'	i	σ_v	σ_v'	σ_d	σ_d'	C_2^{AC}	C_2^{BD}	C_2^{MP}	C_2^{NQ}	I	C_4	C_4'	C_2^z
S_4	S_4	i	σ_h	S_4'	σ_d	σ_d'	σ_v'	σ_v	C_2^{MP}	C_2^{NQ}	C_2^{BD}	C_2^{AC}	C_4	C_2^z	I	C_4'
S_4'	S_4'	σ_h	i	S_4	σ_d'	σ_d	σ_v	σ_v'	C_2^{NQ}	C_2^{MP}	C_2^{AC}	C_2^{BD}	C_4'	I	C_2^z	C_4
i	i	S_4'	S_4	σ_h	σ_v'	σ_v	σ_d'	σ_d	C_2^{BD}	C_2^{AC}	C_2^{NQ}	C_2^{MP}	C_2^z	C_4'	C_4	I

$$\begin{array}{c|c} D_{4h} & S_a \\ \hline S_b & (S_b S_a) \end{array}$$

interest. Thus, considering the symmetry operations that can be performed on $PtCl_4{}^{2-}$ (Fig. 3.9), the C_4 operation taken twice produces the identical configuration obtained by taking the counterclockwise or C_4' operation twice. Furthermore, rotation around the same axis by 180°, C_2'', again gives an identical orientation, and hence we may write $C_4{}^2 \equiv C_4'{}^2 \equiv C_2''$.

In the case of BF_3 (Fig. 3.2) the complete set of symmetry operations is I, C_3, $C_3{}^2$, C_2, C_2', C_2'', σ_v, σ_v', σ_v'', σ_h, S_3, and $S_3{}^2$. Here, a counterclockwise operation C_3' produces the identical orientation obtained from $C_3{}^2$, and $S_3{}^2$ is identical with S_3', and hence S_3' and C_3' are not specified as symmetry operations. C_3 and C_3' are similar symmetry operations, differing only in the direction of rotation, and consequently they are said to belong to one *class*; similarly, S_3 and S_3' belong to one class. The elements σ_v, σ_v', and σ_v'' (and also the elements C_2, C_2', and C_2''), are transformed into one another by application of the operations C_3 and C_3', and such elements are also referred to as belonging to a class. In listing all symmetry elements (or operations) it is sufficient to list one characteristic element of each class, preceded by the number of elements in the class. Consequently, the list of symmetry operations may be given as I, $2C_3$, $3C_2$, $3\sigma_v$, σ_h, and $2S_3$ (six classes, two each of orders 1, 2, and 3).

A little further examination of these multiplication tables is of interest. Referring back to Table 3.1, which respresents the fourth-order group C_{2h}, it will be seen that the first two rows of the first two columns of this table give:

	I	$C_2{}^z$
I	I	$C_2{}^z$
$C_2{}^z$	$C_2{}^z$	I

By our rule this is also a group (we see below that it is the group C_2); it is called a *subgroup* of the entire group of Table 3.1 and is of order 2. I and $\sigma_v{}^{xy}$ or I and i are further subgroups or order 2. Similarly in Table 3.2, I and any one other element form a subgroup of order 2. There are no other subgroups in these two groups.

Examination of Table 3.3, however, shows a tremendous number of subgroups. Thus I together with any of the other elements except C_4, C_4', S_4, or S_4' is a subgroup of order 2. The first four rows of the first four columns are a subgroup of order 4 (call this **A**). The first and last four rows of the first and last four columns are a subgroup of order 8 (call this **B**). Note, however, that the last four rows of the last four columns are *not* a subgroup since the products are C_4, C_4', and C_2^z, which do not occur in the respective column and row headings. But **A** is a subgroup of **B**. Also, the groups of Tables 3.1 and 3.2 are subgroups of the group of Table 3.3. Alternately, we may call the group in Table 3.3 a *supergroup* of all the others.

3.5 POINT GROUPS

We have discussed a number of molecules (*trans*-dichloroethylene, water, BF_3, and $PtCl_4^{2-}$) in terms of the complete set of symmetry operations that make up the point group appropriate to the molecule. Every molecule can be assigned to a point group depending on the applicable set of symmetry elements, and although there is theoretically an infinite number of point groups, these can be classified into a very few types. The point groups of interest in molecular structure work are discussed below.

In listing the various point groups, two systems of notation are again given. The first is based on the *C-σ-i* notation of symmetry elements and is called the Schönfliess notation. The other, shown in parentheses, is that used by crystallographers. It is based on the Hermann-Mauguin system and called the International System, since it is the one adopted and recommended by the International Union of Crystallographers. It is based on enumerating the minimum number of symmetry elements sufficient to define the point group. Thus the symmetry of a molecule like *trans*-dichloroethylene is completely described by $2/m$. since the other element, i, is a necessary result of the presence of these two. The fact that the mirror plane is a horizontal plane, that is, perpendicular to the axis, is indicated by the slash, which is alternately written as $\frac{2}{m}$. Similarly, the water molecule is *mm* and the 2 is implied. *2m* would be equally sufficient (the *m* following the 2 indicating that plane and axis are parallel), but *mm* is usually preferred. Some authors write more than the minimum number of elements, and *2mm* is acceptable. Since $\bar{6}$ is identical to $3/m$, it is often used in this way.

To further illustrate the symmetry elements present in each point group, we show the stereographic projections for a large number of point groups. After the description of each point group, the reader should consult the corresponding stereographic projection in order to fix in his mind the symmetry elements present.

Type 1. No rotational axis; point groups C_1, C_s, C_i.

(*a*) $C_1(1)$. This group has no elements of symmetry (asymmetric compounds). For examples see Fig. 3.10*a*.

(*b*) $C_s(\bar{2})$. This group has only a single plane of symmetry, σ, or $S_1(\bar{2})$. For examples see Fig. 3.10*b*.

(*c*) $C_i(\bar{1})$. This group has only a center of inversion, i, or $S_2(\bar{1})$. For examples see Fig. 3.10*c*.

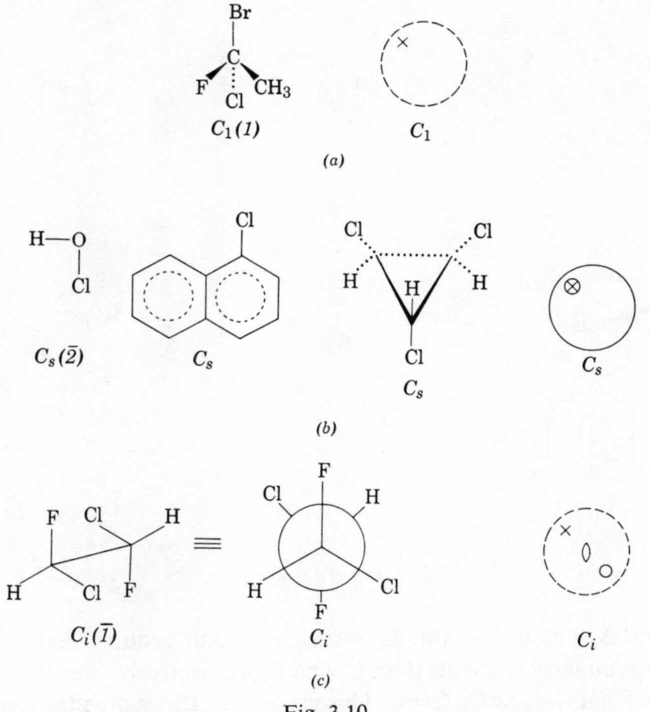

(*a*)

(*b*)

(*c*)

Fig. 3.10

Type 2. Only one axis of rotation; point groups C_p, S_p, C_{pv}, C_{ph}.

(*a*) *Point Groups* $C_p(p)$. This group has only a single rotational axis of order greater than one. These molecules are dissymmetric, and are optically active unless various conformations are readily interconvertible. See Fig. 3.11a.

(b) *Point Groups* S_p, for example, $S_4(\bar{4})$ and $S_6(\bar{3})$. A molecule with S_1 also has $\sigma(\bar{2})$; one with S_2 also has $i(\bar{1})$. Although S_4 also has

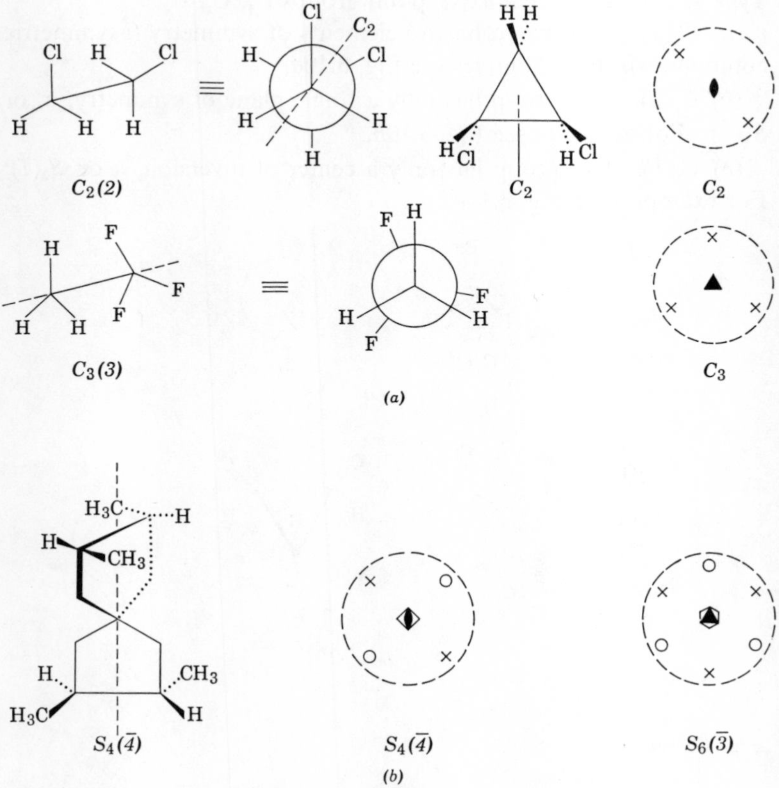

$C_2(2)$ C_2 C_2

$C_3(3)$ C_3

(a)

$S_4(\bar{4})$ $S_4(\bar{4})$ $S_6(\bar{3})$

(b)

Fig. 3.11(a, b)

C_2 and S_6 has C_3, S_4 and S_6 are higher point groups, that is, have more symmetry elements than C_2 and C_3, respectively. See Fig. 3.11b.

(c) *Point Groups* $C_{pv}(pm)$. This group has the symmetry elements C_p and p vertical planes (σ_v) intersecting in the axis. Molecules in these point groups are very common. See Fig. 3.11c.

(d) *Point groups* $C_{ph}(p/m)$. This group has the symmetry element

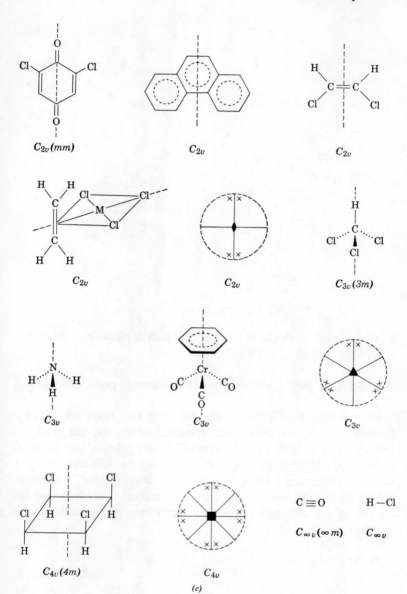

Fig. 3.11(c)

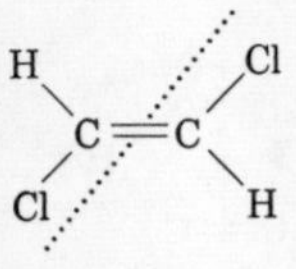

$C_{2h}(2/m)$ 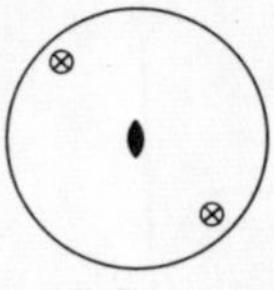 C_{2h} 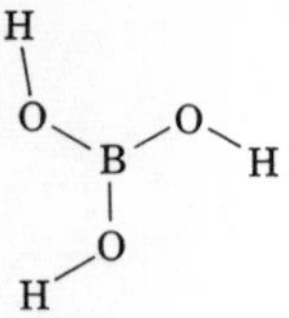 $C_{3h}(3/m)$

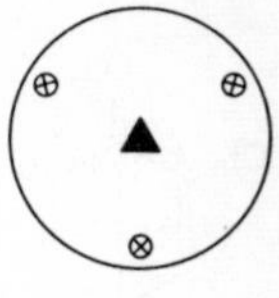

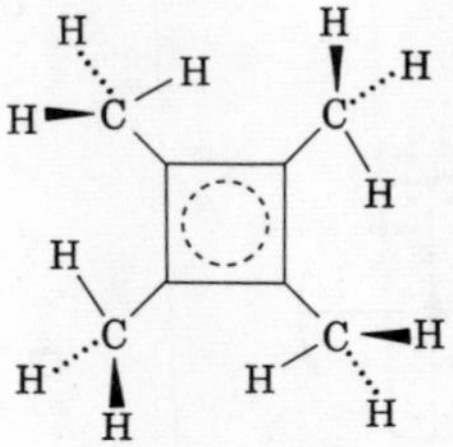

$C_{4h}(4/m)$ 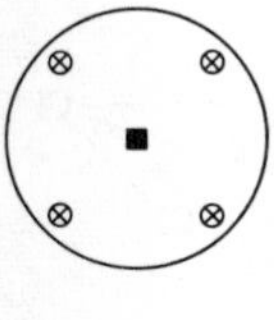 C_{4h}

(d)

Fig. 3.11(d)

C_p and at right angles to it, a horizontal mirror plane, σ_h. When p is even, a σ_h implies i. See Fig. 3.11d.

Type 3. One p-fold axis and p-twofold axes; point groups D_p, D_{ph}, D_{pd}.

Point Groups D_p. If, to C_p, the principal axis, there are added p twofold axes as the only other symmetry elements, the group is $D_p(p2)$. The D is from the German "Diedergruppe," or dihedral group. D_2 (222), which has three equivalent twofold axes at right angles, is also called V from the German "Vierergruppe." Molecules belonging to this point group are not common. Ethylene and biphenyl in an orientation neither planar nor perpendicular are

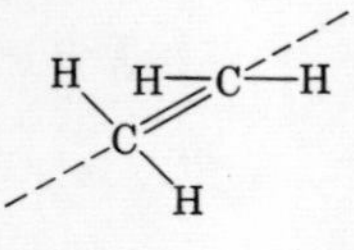

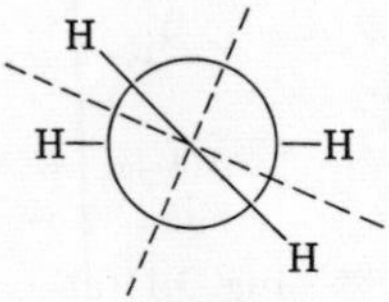

$D_2(222)$ 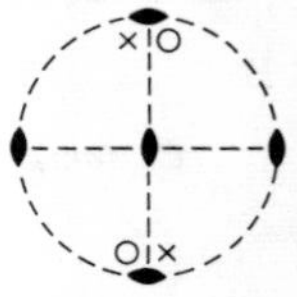 D_2

Fig. 3.12

D_2 (222; here, because they are all 2's, all three axes are written). Usually one axis in this group is quite obvious. The other two axes are in a plane at right angles to this axis and are at right angles to each other. For examples see Fig. 3.12.

Point Group D_{ph} [$p/m\ 2/m\ 2/m$ or $(p/m)mm$]. If σ_h is added to the C_p axis and p twofold axes, the molecule belongs to this point group. The addition of σ_h results in $p\sigma_v$. In a specific example this can be most easily seen by examination of the stereographic projection for D_2 above, where addition of σ_h immediately means that the two two-fold axes now also define two σ_v planes, each including one C_2 axis and the C_p vertical axis. Furthermore, when p is an even number, a center of inversion is also implied. The case of D_{2h} is special because none of the three twofold axes stands out over the others. The special designation V_h is sometimes used for this point group since as we saw above $V = D_2$. Many molecules possess D_{ph} symmetry. For examples see Fig. 3.13.

Point Group D_{pd} ($\bar{p}2m$, where $\bar{p}$ is the appropriate rotation-inversion axis, not the same as the p in D_p). If besides the axes defining D_p there are p diagonal planes σ_d which bisect the angles between successive twofold axes, the molecule belongs to this point group. Thus D_{2d} would have the symmetry elements $3C_2$ and $2\sigma_d$ and these elements also imply a fourfold rotation-reflection axis, S_4. Many common molecules consisting of two equal halves at right angles to one another belong to this point group. In the general case there would be in D_{pd} an S_{2p} axis, and if p in D_{pd} is odd, there is also a center of inversion. See Fig. 3.14.

Type 4. More than one axis higher than twofold; point groups T_d, O_h; also T, T_h, O, I, I_h, and K_h.

(a) *Point Group* $T_d(\bar{4}\ 3\ m)$. The common tetrahedral molecules with identical groups around a central atom belong to this point group. The symmetry elements present are $4C_3$, $3C_2$ (which are also S_4), and 6σ, corresponding to the operations $8C_3$ (that is, $4C_3$ and $4C_3'$), $3C_2$, $6S_4$ ($3S_4$ and $3S_4'$), and 6σ. The axes through the center and a corner (vertix) of the tetrahedron are the C_3 axes; since there are four corners (vertices) there are four C_3. The axes through the midpoints of opposite pairs of edges are the C_2 axes; since there are six edges, there are three C_2 axes. Each edge lies in one mirror plane, and since there are six edges, there are six σ. For examples see Fig. 3.15.

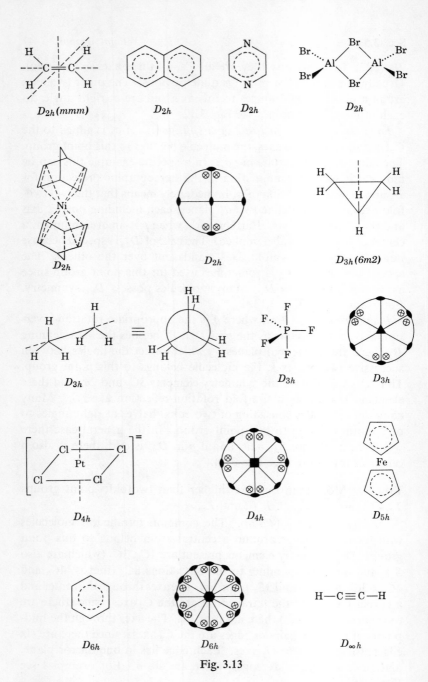

Fig. 3.13

48

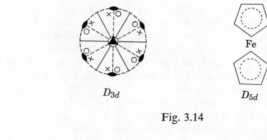

D_{2d} or $V_d(\bar{4}2\,m)$ D_{2d} D_{2d}

$D_{3d}(\bar{6}2\,m)$ D_{3d}

D_{3d} D_{5d}

Fig. 3.14

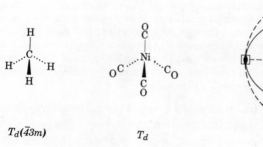

$T_d(\bar{4}3m)$ T_d T_d

Fig. 3.15

At this point it becomes necessary to discuss a little further the stereographic representations of axes and planes that are neither horizontal nor vertical, such as are encountered in the point group T_d. If the methane molecule is oriented so that one of its S_4 axes is vertical, the other two in the x and y axes (Fig. 3.16a) result. Each of the CH bonds defines a C_3 axis, as is readily seen for one such bond in Fig. 3.16b. If the molecule in the orientation shown in Fig. 3.16a is inscribed in a sphere—to produce a spherical projection—these four C_3 axes intersect the sphere in eight points, points which are the corners of a cube inscribed in the sphere. If the corners of this cube are now projected stereographically on the equatorial plane—that is, by drawing the line from the corner

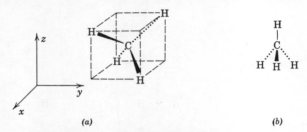

(a)   (b)

Fig. 3.16   The CH$_4$ molecule of point group T_d.

to the opposing (north or south) pole—four pairs of points arise. Each pair corresponds to a corner above and one below the equatorial plane. These pairs of points are indicated by the symbol (triangle) for the C_3 axis and are connected by broken lines. In the point group T, having no planes, these, together with the symbols for the three S_4, would complete the stereographic projection.

T_d, however, to which the methane molecule belongs, in addition to these axes possesses six symmetry planes, each containing the C atom and two H atoms. Two of these, the one involving the upper two H's of Fig. 3.16a and the one involving the lower ones, are vertical, and accordingly are indicated by solid, straight lines through the triangles representing the threefold axes. The other four planes are tilted. Each of the tilted planes cuts the surrounding sphere in a great circle, going through four corners of the inscribed cube. Projection of these great circles stereographically on the equatorial plane leads to ovals which touch the equatorial great circle at two points each, the points where the tilted great circle passes through the equatorial plane. Pairs of planes have the same intersection with the equatorial plane, one tilted to one side, the other to the other side. In the stereographic projection, pairs of such planes are indicated by a single oval solid line. These symmetry elements are shown in the stereographic projection for T_d, Fig. 3.15.

(b) Point Group O_h ($m3m$ or $4/m\,\overline{3}\,2/m$). This is the point group to which all symmetrical octahedral molecules belong. The cube also belongs to this point group; many of the symmetry elements are most readily recognized in this common geometrical object. The

symmetry elements are $3C_4$ (operations $3C_4$ and $3C_4'$), $4C_3$ (operations $4C_3$ and $4C_3'$), $6C_2$ (other than the C_4), i, and 9σ. The C_4 axes pass through the center of pairs of opposite faces of the cube, and since there are six faces there are three C_4. The C_3 axes pass through pairs of opposite corners and the center of the cube, and since there are eight corners there are four C_3. The C_2 axes pass through pairs of opposite edges as well as the center of the cube, and since there

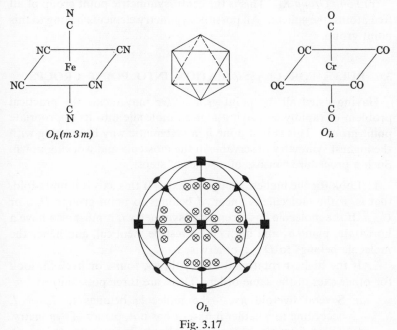

Fig. 3.17

are twelve edges, there are six C_2. Some mirror planes bisect opposite faces (six faces = three σ), and others opposite edges (twelve edges = six σ), and hence there are nine σ. (For examples see Fig. 3.17.)

Other point groups such as $T(23)$, which is T_d without the planes; T_h ($m3$ or $2/m\ 3$), which is T plus i; O (43 or 432), which is O_h without the center and planes; I (53 or 532), possessing $6C_5$, $10C_3$, $15C_2$; and $I_h(5/m\ \bar{3}\ 2/m)$, which has I plus i, additional planes, and S axes, have all been omitted because they are of minor importance in dealing with chemical compounds. The first example of a molecule belonging to I_h, dodecaborane ion, $B_{12}H_{12}{}^{2-}$, has recently been reported. This unusual ion, one of the most symmetrical known, has twelve hydrogen atoms and twelve boron atoms. Its geometry is that of an icosahedron with a boron atom at each vertex of the twenty equilateral triangles that are its faces. The ion has the following symmetry elements: 15σ, $6C_5$, $10C_3$, and at least $15C_2$. All told, there are 120 possible symmetry operations.

(c) *Point Group K_h.* This is the centrosymmetric point group of all free atoms, the sphere. All possible symmetry elements belong to this point group.

3.6 RULES FOR CLASSIFICATION INTO POINT GROUPS

Having listed all the point groups, we may tackle the practical problem of rapidly classifying a given molecule into its appropriate point group. This is best done in a systematic way by starting with the highest symmetry observable in the molecule and working down. Such a procedure consists of a series of steps.

1. Look for the highest rotational axis. If this axis is infinite-fold, that is, if the molecule is linear, it belongs to point groups $D_{\infty h}$ or $C_{\infty v}$. If the molecule has a center of symmetry i, it must also have a horizontal plane σ_h and both ends must be identical, and hence the molecule belongs to $D_{\infty h}$; otherwise to $C_{\infty v}$.

2. If the highest rotational axis is three-, four-, or fivefold, look for other axes of the same order. There are three possibilities:

   a. Several fivefold axes—the molecule belongs to I_h or I, according to whether it has, or has not, planes of symmetry.
   b. Three fourfold axes—the molecule belongs to O_h or O, according to whether it has, or has not, planes of symmetry.
   c. Four threefold axes, but no four- or fivefold axes. In the absence of a center and planes of symmetry, point group T is indicated; with a center, point group T_h; and with *six* planes and three S_4 axes, T_d.

3. If only one axis has $p > 2$, or if the axis of highest order is twofold, check for p more twofold axes at right angles. If these exist, the molecule belongs to D_p if it has no planes, to D_{ph} if it has a horizontal plane, and to D_{pd} if it has p vertical planes but no horizontal one.

4. If only a single p-fold axis exists, check for an S_{2p}. If this exists, it is the point group; if not, the molecule belongs to C_p if it has no planes, to C_{ph} if it has one (horizontal) plane, and to C_{pv} if it has p (vertical) planes.

5. If no axis exists, a plane indicates the point group C_s, a center C_i, and none of these elements C_1.

The problem often arises which of two molecules, or of two point groups, is more symmetrical. This question does not always have an answer. It is obvious that C_1 has lower symmetry than C_2, C_s, or C_i, or for that matter, than any other point group. But which is the most symmetric, C_2, C_s, or C_i? All three imply a single element of symmetry, all general atoms occur in equivalent pairs, and only those on the element are unique. In C_s there may be more unique atoms, since σ is a two-dimensional element; C_i can only have a single one, since i is a point. But in spite of this, the question has no answer. The problem is similar for C_{ph} and C_{pv}, for D_{pd} and D_{ph}. Thus, a point group (for example, C_{pv}) formed from another one, C_p, by *addition* of one or more symmetry elements, called a supergroup of the simpler one, is *more* symmetric and has *higher* symmetry, but unless there is such a generic relation, a comparison is not valid. Similarly, a group like T, which has some but not all of the symmetry elements of another group, say T_d, but no others, is called a *subgroup*, and is *less* symmetric. Figure 3.18 shows the hierarchy of some of the types of point groups and serves particularly to emphasize their interlacing.

Let us come back to the matter of subgroups for a moment. Take again D_{4h}, for which the multiplication table is given in Table 3.3, and let us try to find the subgroups. We will only give the numbers of the columns (together with the equally numbered rows) and the group: 1,4 (or 1,5; 1,6; 1,7; 1,8)C_2; 1,9 (or 1,10; 1,11; 1,12; 1,13)C_s; 1,16, $S_2 = C_i$; 1,2,3,4,C_4; 1,4,5,6 (or 1,4,7,8)D_2; 1,4,9,10 (or 1,4,11,12; 1,5,9,13; 1,6,10,13; 1,7,11,13; 1,8,12,13)C_{2v}; 1,4,13,16 (or 1,5,10,16; 1,6,9,16; 1,7,12,16; 1,8,11,16)C_{2h}; 1,4,5,6,-11,12,14,15,16, and others, D_{2d}; 1,4,5,6,9,10,13,16 and others, D_{2h}; 1,2,3,4,5,6,7,8,D_4; 1,2,3,4,9,10,11,12,C_{4v}; 1,2,3,4,13,14,15,16,C_{4h}. This indicates to quite a degree the matter of hierarchy: We can start with C_1 and add elements successively to generate supergroups as shown in Fig. 3.18.

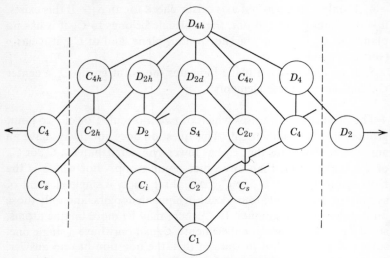

Fig. 3.18   Hierarchy of symmetry groups.

3.7   EQUIVALENT ATOMS

A further problem of considerable interest which will concern us in connection with normal vibrations in Chapter 5 has been briefly mentioned: How many atoms in a molecule belonging to a given point group are equivalent, or, how many times must any given atom be repeated in the molecule? This question is readily settled by reference to stereographic projections. Since these projections show all equivalent positions of a *general* point, it is sufficient to count the number of times the symbols (cross or circle) appear, to determine the number of equivalent *general* atoms, that is, atoms not lying on any symmetry element. Thus, from the figures given it is obvious that this number is 1 for C_1, 2 for C_s, C_i, or C_2, 3 for C_3; 4 for C_{2h}, C_{2v}, or D_2; and 24 for T_d and O_h. For other atoms, the determination requires placing a point on the symmetry element in question, and performing the operations. Thus, in C_p, C_s, or C_i, an atom lying on the element is unique, as is, in other point groups, any atom lying on *all* elements (for this purpose, for an S axis or a rotation-inversion axis, the element is the center of gravity). In D_2, C_{2v}, or C_{2h}, a point lying on one $C_2(D_2)$, on one plane (C_{2v}), or on either the C_2 or the plane (C_{2h}), but not on the other elements, must occur twice, as may be readily verified.

The equivalences discussed in the preceding paragraph are *symmetry* equivalences. There may be, however, additional *chemical* equivalences. Take, for example, trichloroacetic acid, Cl_3CCOOH. The C—CCl_3 group by itself has a C_3 axis and three σ_v planes, and hence C_{3v} symmetry; this is called the *local symmetry* of this group.

The $C{-}C\overset{\displaystyle O}{\underset{\displaystyle OH}{\Big/\Big\backslash}}$ group, however, is normally planar, and has local

symmetry of C_s. The total symmetry of the molecule, then, is C_s in the conformations in which one Cl lies in the COOH plane, and C_1 (that is, asymmetric) in all other conformations. The three Cl atoms, although symmetrically nonequivalent, are, however, in virtually identical environments, and are consequently chemically equivalent, even without free rotation about the C—C bond. This rotation, of course, assures their complete equivalence. This equivalence is reflected in a statistical factor of three, applicable to any theoretical treatment of the thermodynamics or kinetics of any reaction of one of the Cl groups. This type of statistical factor is particularly important in the treatment of the pK's of polycarboxylic acids.

In nuclear quadrupole spectroscopy, which is performed in the solid state, the problem is reversed. Here, conformations in the *solid* are generally fixed. In addition, the *molecular* symmetry, as will be seen in Chapter 6, is not necessarily the same as the crystal symmetry. Consequently, a given pair of atoms may be equivalent in the molecule, but need not be equivalent in the solid; hence the common observation in NQR spectra of chemically, and even symmetrically (in the molecule) equivalent atoms showing slightly different resonances. Thus there are sixteen distinct crystal sites for chlorine atoms in CCl_4. Fifteen distinct resonances are observed; the sixteenth seems to be superimposed on one of the others.

A final problem concerning the conformations of molecules arises in connection with the *symmetry number* appearing in the calculation of the rotational partition function of statistical mechanics. This measures the number of equivalent positions successively attained in a complete rotation about a given axis, and is, of course, just the order of the axis.

REFERENCES

E. P. Wigner, *Group Theory and Its Applications to Quantum Mechanics of Atomic Spectra*, Academic Press, New York, 1959.

H. Boerner, *Group Representations*, Springer, Berlin, 1955.

M. Hamermesh, *Group Theory and Its Application to Physical Problems*, Addison-Wesley, Reading, Mass., 1962.

H. Weyl, *The Classical Groups*, Princeton University Press, Princeton, 1946.

See also the References at the end of Chapter 2.

4

Group Theory

....

In Chapters 1–3 we have discussed the symmetry properties of molecules (and, by implication, of any solid body). Such symmetry behavior is appropriate only to the stationary molecule. In the present chapter we shall examine how various physical and mechanical properties of molecules behave under symmetry operations. Such properties can readily be classified into two major types, with possible extensions. The simplest are scalar, that is, properties which have a value but no direction, such as mass, volume, and temperature. These properties obviously do not depend in any way on symmetry properties; for example, the mass of a molecule remains the same when the molecule is rotated by 180° about some axis; it does not even matter whether or not the axis is a symmetry axis and, if it is, whether it is twofold, threefold, or whatever its order.

Other properties of molecules, however, have not only magnitude but also direction, and their behavior under symmetry operations is more complicated. Imagine, for instance, that you are standing in front of a mirror and throwing a ball up and parallel with it; the mirror image will be moving right along with the ball and at the same speed. Now throw the ball directly (perpendicularly) at the mirror—the image will appear to be moving in a direction opposite to that of the ball and the image and the ball collide head-on at the mirror; both the ball and its image again travel at the same speed. These two types of behavior are called *symmetric* and *antisymmetric*, respectively, and are characteristic of typical directional, so-called vectorial properties of matter.

4.1 NONDEGENERATE POINT GROUPS

Translational Motion. Let us examine the motion of a typical molecule, for example, water, in the direction of each of the Cartesian axes under the various symmetry operations. For H_2O, which belongs to the point group C_{2v}, there are four of these: I, C_2^z, σ_v^{xz}, and σ_v^{yz}. Now take the motion in the y direction, that is, to the right, as in Fig. 4.1a. Reflection on σ_v^{xz} can be visualized by regarding the motion of the image in a mirror parallel to the xz plane, but to the right of the molecule. This mirror image will be moving in the *opposite direction*, although with the same speed. In other words, the

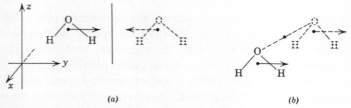

(a)   (b)

Fig. 4.1 The reflection of H_2O in a mirror (a) parallel the xz plane and (b) parallel the yz plane.

direction of the motion is reversed by reflection at any plane parallel to xz, and including the xz plane itself. This behavior is given expression by saying that the motion is *antisymmetric* with respect to the reflection. If, on the other hand, we observe the same motion in the y direction of the original water molecule in a mirror *behind* the molecule, that is, parallel to the yz plane (cf. Fig. 4.1b), we find that the mirror image moves in the *same* direction as the molecule. The motion is *symmetric* with respect to reflection in the yz plane. Of course the mirrors are not needed; they only help us to visualize the behavior under the symmetry operation. If we apply the C_2 operation to this same motion in the y direction, the motion to the right reverses into motion to the left, and is again antisymmetric with respect to rotation about C_2^z. The identity operation, of course, leaves the motion unaffected.

   If we observe motion in the z direction (upward) in the two vertical mirrors (parallel to the two σ_v) the mirror images move along with the molecule, the motion in the z direction is symmetric with respect to both σ_v. Motion in the z direction is also symmetric with respect to I and $C_2{}^z$. Finally, motion in the x direction is found to be symmetric with respect to I and $\sigma_v{}^{xz}$, antisymmetric with

Table 4.1　Character Table for Point Group C_{2v}

		I	$C_2{}^z$	$\sigma_v{}^{xz}$	$\sigma_v{}^{yz}$	Designation	
	z	+1	+1	+1	+1	A_1	Γ_1
Translation	x	+1	−1	+1	−1	B_1	Γ_3
parallel	y	+1	−1	−1	+1	B_2	Γ_4
	z	+1	+1	−1	−1	A_2	Γ_2
Rotation	y	+1	−1	+1	−1	B_1	Γ_3
about	x	+1	−1	−1	+1	B_2	Γ_4

respect to $C_2{}^z$ and $\sigma_v{}^{yz}$. All these facts may be given formal expression by the factors shown in Table 4.1, where $+1$ refers to symmetric and -1 to antisymmetric behavior, for reasons which will become obvious shortly.

Rotational Motion.　Let us now examine the similar transformations of rotational motion about the three Cartesian coordinates, considering first rotation around the z axis (Fig. 4.2). Such rotation will

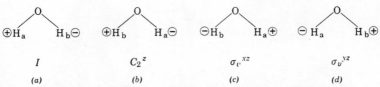

Fig. 4.2　The counterclockwise rotation of the water molecule about the z axis. A circled plus indicates motion upward, out of the plane of the paper, the yz plane; a circled minus indicates motion downward, into this plane. (*a*) The original molecule, and the molecule resulting from I, identical to the original. (*b*) The molecule resulting from $C_2{}^z$, (*c*) from $\sigma_v{}^{xz}$, and (*d*) from $\sigma_v{}^{yz}$.

take one of the hydrogen atoms forward out of the plane, while the other hydrogen is moving backward out of the plane and the oxygen atom remains in the plane. The motion toward the observer is denoted by a circled plus, and the motion away from the observer by a circled minus, as if an arrow were coming toward the observer in the "plus" case and receding from the observer in the "minus" case.

Performing the identity operation I on the molecule while rotating the molecule in a counterclockwise direction around z leaves the rotational motion unchanged. Performing the C_2^z operation on the counterclockwise rotating molecule (Fig. 4.2a) interchanges H_a and H_b, but the H atom on the left will still be moving toward and the H atom on the right away from the observer. The resulting molecule is shown in Fig. 4.2b. If (a) is submitted to the σ_v^{xz} operation, that is, reflection in the plane perpendicular to the plane of the paper, H_a and H_b as well as their directions of motion are interchanged, and the orientation shown in Fig. 4.2c results. Thus H_b, which was moving away in the original, is now, after the operation σ_v^{xz}, moving toward the observer, and similarly H_a has also changed direction. When Fig. 4.2a undergoes the σ_v^{yz} operation, H_a on the left and moving out remains H_a but moves away (arrowhead away from the observer). The direction of rotation about the z axis is thus unchanged by I and C_2^z, reversed by σ_v^{xz} and σ_v^{yz}. Similarly, Figs. 4.3 and 4.4 show the results of operating with C_2^z, σ_v^{xz}, and σ_v^{yz} on the rotations about the x and y axes, respectively. These relations are represented by the last three rows of Table 4.1.

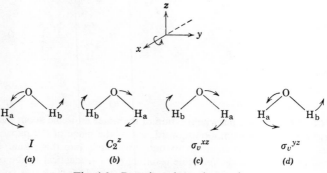

Fig. 4.3   Rotation about the x axis.

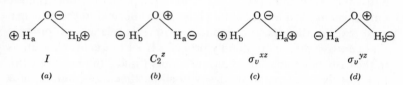

Fig. 4.4   Rotation about the y axis.

Symmetry Species.   Inspection of Table 4.1 shows that the behavior of the three motions in the x, y, and z directions (translational motions) and the rotations about the z axis fall into four different behavior patterns.   However, the rotations about the x and y axes behave just like the translational motions in the y and x directions, respectively.   Thus we have found four distinct behavior patterns which we shall call *symmetry types*, or, more commonly, *symmetry species* or *irreducible representations*.   Do these four types exhaust the possible patterns? I must always be represented by $+1$,[1] but it would seem that the possible assignments of $+1$ or -1 to the three other operations would yield 2^3 or 8 types.   It has been shown in Chapter 3 that in the point group C_{2v} only two of the nontrivial operations (operations other than I) are independent, while the third (and I) arise from combinations (multiplication) of the other two.   Since there are just four (2^2) ways to assign $+1$ and -1 to *two* independent operations, there are just four symmetry species, the four which we have found.

The symmetry species have, for convenience, been given shorthand symbols, which are given in the last two columns of Table 4.1.   By convention, the species that are symmetric with respect to the rotational axis are designated by A, those antisymmetric by B.   If several axes exist, the one of highest order determines the A–B symbol.   If several axes of the highest order exist, A refers to species symmetric

---

[1] This is true only in the point groups not involving axes above twofold.   We return to this point in discussing degenerate species.

with respect to *all* of them, *B* to species antisymmetric with respect to *any* of them. Where more than one *A* or *B* species exist, they are distinguished by subscripts (or sometimes primes). If a center of symmetry is present, the subscript *g* (for *gerade*, German for even) is used for species symmetric, *u* (for *ungerade*, odd) for species antisymmetric with respect to the center. If a plane of symmetry is the only element, or if a horizontal plane (plane normal to the principal axis) exists, the species symmetric to this plane are primed; those antisymmetric are doubly primed. If all this is insufficient, as in the case under consideration, subscripts 1 and 2 (and 3) are used. In the *A* series, A_1 is the species with all positive signs, but in the *B* species the assignment of 1 and 2 is arbitrary.

. In an alternate system, each species or each representation is given the symbol Γ (capital gamma). The different representations are distinguished by number subscripts on Γ in the logical order; first all species *A*, in order of increasing number of subscripts or primes, then the *B*, and then the *E* and *F* or *T*, which will be introduced later. If *g* and *u* classification applies, *g* is numbered before *u*.

Molecular Orbitals. We have shown thus far how two extremely simple nonstationary properties of molecules—translational and rotational motion—may be classified in symmetry species. The same type of classification is possible for other properties. We here discuss two more types: vibrational motion, that is, the motion of various atoms within the molecule relative to each other; and electronic motion, that is, the electronic wave functions which describe, according to quantum mechanics, the property classically described as motion of electrons.

The wave functions are generally most simply treated by a discussion of the areas in which the function is positive or negative. We will use the usual approximation that a separate wave function can be written for each electron (orbital approximation). According to quantum mechanics, the square of the wave function describes the probability of finding an electron in a given place and hence is a stationary, scalar property of a molecule. Consequently, it must have the same symmetry properties as the molecule itself; that is, it must transform into itself under each of the applicable symmetry operations. (This is true only in nondegenerate groups; we will return to the others later.) This is possible, and possible only, if each wave function transforms as one of the symmetry species of the point group to which the molecule belongs.

To illustrate, we give schematic representations of the occupied molecular orbitals of water. In the LCAO (linear combination of

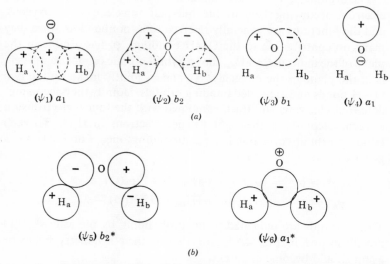

$(\psi_1)\,a_1$ $(\psi_2)\,b_2$ $(\psi_3)\,b_1$ $(\psi_4)\,a_1$

(a)

$(\psi_5)\,b_2{}^*$ $(\psi_6)\,a_1{}^*$

(b)

Fig. 4.5 (a) The four occupied MO's of H_2O, ψ_1–ψ_4, and the symmetry species to which they belong. (b) The antibonding MO's of H_2O, ψ_5 and ψ_6.

atomic orbitals) approximation, these are approximately given by the forms:

$$
\begin{aligned}
\psi_1 &= a_1 = 1s(H_a) + 1s(H_b) + \lambda_1 2s(0) + \lambda_2 2p_z(0)\\
\psi_2 &= b_2 = 1s(H_a) - 1s(H_b) + \lambda' 2p_y(0)\\
\psi_3 &= b_1 = 2p_x(0)\\
\psi_4 &= a_1 = 2s(0) - \lambda'' 2p_z(0)
\end{aligned}
\tag{4.1}
$$

These orbitals are schematically represented in Fig. 4.5a. Operation on each of these four functions with each of the four symmetry operations—which is left to the reader—readily shows that each transforms as the symmetry species indicated in Eq. (4.1) and Fig. 4.5a. We should add parenthetically that it is customary to represent orbitals like those shown in Fig. 4.5a and Eq. (4.1), which are functions of one electron only, by lower-case symmetry symbols, and many-electron wave functions, or determinants of many electron functions, by capital letters. The symmetry symbol is often used to stand for the wave function, sometimes modified by prefixing a "quantum number."

The orbitals ψ_3 and ψ_4 are functions of the oxygen atom only (the electrons occupying them are the lone-pair or n electrons on oxygen), and the functions are generally referred to as nonbonding since they make no contribution to the O—H bonding. ψ_3 has a node in the plane of the molecule, and ψ_4, the sp hybrid atomic orbital on oxygen, is in the plane of the molecule. ψ_1 and ψ_2 jointly make up the two O—H bonds and are called bonding orbitals. Jointly they accommodate four electrons and thus, combined with the four n electrons on oxygen, account for the eight valence electrons in H_2O. To each bonding orbital corresponds an antibonding one. For H_2O this is given by

$$\psi_5{}^* = b_2{}^* = 1s(H_a) - 1s(H_b) - \lambda'''2p_y(O)$$
$$\psi_6{}^* = a_1{}^* = 1s(H_a) + 1s(H_b) - \lambda_1''''2s(O) - \lambda_2''''2p_z(O) \tag{4.2}$$

These orbitals are indicated to be antibonding by the asterisk, and are illustrated in Fig. 4.5*b*, from which their symmetry behavior again should become apparent.

At this point, it becomes convenient to give a preliminary interpretation of the meaning of the $+1$ and -1 (the so-called character) given in Table 4.1. The wave function is actually a complicated mathematical function which has some numerical value in every point in space. We do not have to worry about the magnitude of this value, only its sign. When the behavior with respect to some *one* operation is antisymmetric, this means that the sign changes, though the absolute value remains unchanged. This change of sign is achieved by multiplication by -1, and thus the entries in Table 4.1 are just factors by which the wave function prior to the operation is multiplied to obtain the wave function resulting from the operation. The factor $+1$ obviously represents symmetric behavior adequately, since it, like the operation, leaves the wave function unaffected.

Vectorial Properties. It is profitable now to examine the translational and rotational motions a little more closely, with greater mathematical rigor. The translational motion is a vectorial property, that is, a property with direction. The motion itself is not a good quantity—it has no magnitude. It is, however, characterized by either of two well-defined quantities, velocity **v** or momentum **p** (given by $m\mathbf{v}$, where m is the mass of the moving particle).

For the reader unfamiliar with vectors and their algebra, the following paragraphs give an elementary introduction. Vectors are quantities with magnitude and direction, and hence can be graphically represented by an arrow whose

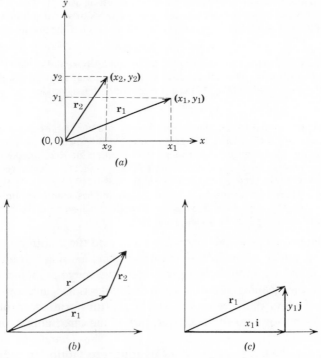

Fig. 4.6 The vector sum **r** of $\mathbf{r}_1$ and $\mathbf{r}_2$.

length reflects the magnitude. Figure 4.6a represents a vector $\mathbf{r}_1$ in the xy plane of some (arbitrary) coordinate system. Its magnitude, r_1, corresponds to the distance from the origin $(0,0)$ to the end of the vector (x_1, y_1). The vector $\mathbf{r}_1$ may be analyzed into two components which are vectors parallel to the x and y axes, of length x_1 and y_1, respectively, which are the projections of $\mathbf{r}_1$ on the x and y axes; its length is $\sqrt{x_1{}^2 + y_1{}^2}$. The sum of two vectors $\mathbf{r}_1$ and $\mathbf{r}_2$ is another vector, **r**, obtained by placing the tail of the second vector, $\mathbf{r}_2$, at the head (arrowpoint) of the first ($\mathbf{r}_1$), and connecting the tail of $\mathbf{r}_1$ with the head of

r_2, as shown in Fig. 4.6b. The length r of $\mathbf{r} = \mathbf{r}_1 + \mathbf{r}_2$ is $r = \sqrt{x^2 + y^2} = \sqrt{(x_1 + x_2)^2 + (y_1 + y_2)^2}$. From this it can be seen that the original vector $\mathbf{r}_1$ can be regarded (Fig. 4.6c) as the sum of two vectors, one along the x axis of length x_1, called $x_1\mathbf{i}$, the other along the y axis of length y_1, called $y_1\mathbf{j}$, where $\mathbf{i}$ and $\mathbf{j}$ are vectors of unit length parallel to the x and y axes, respectively—so-called unit vectors. If our vector $\mathbf{r}_1$ lay in three-dimensional space rather than in the xy plane, $\mathbf{r}_1$ would have components $x_1\mathbf{i}$, $y_1\mathbf{j}$, and $z_1\mathbf{k}$, where $\mathbf{k}$ is a unit vector in the z direction.

The multiplication of two vectors has some interesting aspects. Two different types of product exist—the dot or scalar and the cross or vector product. The dot product is a scalar (quantity without direction) and has the magnitude

$$\mathbf{r}_1 \cdot \mathbf{r}_2 = r_1 \cdot r_2 \cdot \cos \theta$$

where r_1 and r_2 are the absolute magnitudes or lengths of $\mathbf{r}_1$ and $\mathbf{r}_2$, and θ is the angle between them. The dot product of two vectors at right angles thus is zero. The cross product (Fig. 4.7) is a new vector with magnitude

$$|\mathbf{r}_1 \times \mathbf{r}_2| = r_1 \cdot r_2 \cdot \sin \theta$$

and direction normal to the plane defined by $\mathbf{r}_1$ and $\mathbf{r}_2$. Its positive direction is determined by a right-hand rule, that is, if $\mathbf{r}_1$ is along the right thumb and $\mathbf{r}_2$ along the right index finger, the cross product is along the right middle finger. The cross product of two parallel vectors is zero, and the product vector is maximum when the vectors form a right angle.

Multiplication of a vector by a scalar simply implies multiplication of its absolute value by the scalar, leaving the direction unchanged if the scalar is positive, or reversing it if the scalar is negative.

At the beginning of this chapter we analyzed the "motion" of the molecule under symmetry operations. The analysis would be identical had we applied the operations to velocity (or momentum) vectors; the meaning of the factors of Table 4.1 is now clearer—they are the factors multiplying the vector to transform the property before operation into the one after, just as indicated above for the orbitals.

Thus far, we have restricted our treatment to motion in either of the three coordinate directions. The physical motion, however, is not so restricted. Again, the vector notation helps resolve the problem. As we saw, each vector $\mathbf{v}$ in three-dimensional space may be resolved into three components according to:

$$\mathbf{A} = A_x\mathbf{i} + A_y\mathbf{j} + A_z\mathbf{k}$$

where $\mathbf{i}$, $\mathbf{j}$, and $\mathbf{k}$ are unit vectors in the x, y, and z directions.

Under the symmetry operation, each of these unit vectors transforms as outlined above, and the resultant vector is the vector sum of the three transformed components. In this way the behavior under a symmetry operation of a velocity in an arbitrary direction may be obtained.

Angular Momentum. Just as for translational motion, there is an alternate, even more useful description of a rotational motion in terms of a vector quantity called the angular momentum. This is a vector that lies along the direction of the axis of rotation.

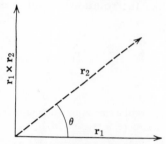

Fig. 4.7 The cross product $r_1 \times r_2$ of the two vectors r_1 and r_2 ($r_1 \times r_2$ is in the plane of the paper, r_1 and r_2 in a plane perpendicular to the paper).

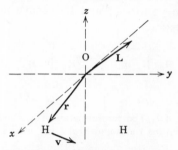

Fig. 4.8 The angular momentum vector (all vectors in the yz plane except L, which lies along the x axis perpendicular to the yz plane).

Linear momentum is a vector in the direction of motion equal to the sum of the products of the mass at any point times the velocity of the point; it reflects the tendency of the object to remain in motion. Angular momentum (L) is the sum of the cross products of linear momentum ($m\mathbf{r}$) and the velocity ($\mathbf{v}$)

$$L = \sum m\mathbf{r} \times \mathbf{v}$$

of a given point during the rotation. The distance (radius vector $\mathbf{r}$) is the vector from the rotational axis to this point, and the sum extends over all mass points. The equation means that L is a vector having length equal to m times the products of the lengths of $\mathbf{v}$ and $\mathbf{r}$, in a direction at right angles to these, the positive direction determined by the right-hand rule. The vectors $\mathbf{v}$ and $\mathbf{r}$ are illustrated for rotation of the water molecule about the x axis in Fig. 4.8. It can be verified that the contributions from each atom to L are in the same direction, and hence we can use the motion of any one of the three atoms to derive the symmetry properties of L.

If either **r** or **v**, but not both, inverts under a symmetry operation, **L** inverts; if both remain unchanged, or both invert, **L** is unchanged. In Fig. 4.9 this is shown for the rotation of water, around the x axis, using the O atom as reference point. If we use the H atoms as reference point, however, we see that both vectors change somewhat in direction without inverting completely; the resulting angular momentum vector, however, behaves the same, whatever the choice of reference point. Similarly, we can show that the angular momentum vectors L^z and L^y behave just like the rotations about these axes.

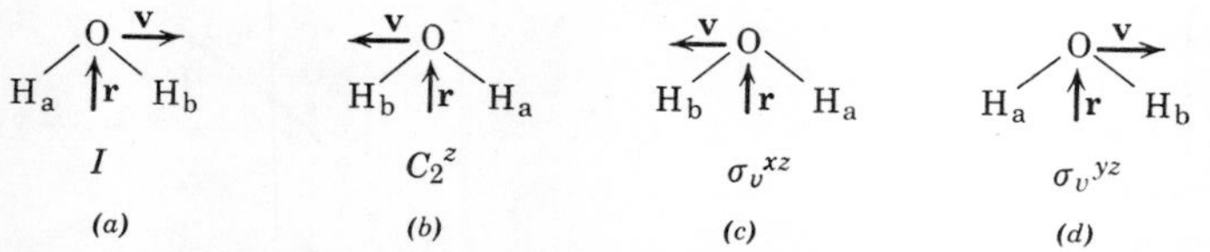

Fig. 4.9   The transformation of the **r** and **v** vectors of water, representing rotation about the x axis under (a) I, the original; (b) $C_2{}^z$; (c) $\sigma_v{}^{xz}$; and (d) $\sigma_v{}^{yz}$.

A velocity vector in an arbitrary direction, can be resolved into its components; similarly, an angular momentum vector in an arbitrary direction, that is, a rotation about an arbitrary axis, can be resolved into its components, and the behavior of an arbitrary rotation can thus be reduced to a vector addition problem.

Dipole Moments. We have now developed the symmetry species and Table 4.1, called a character table, for the water molecule and hence for the point group C_{2v} by consideration of the velocity vectors, that is, the translational motion, and the angular momenta, that is, the rotational motion. We can now look into the implication of these classifications for other vectorial properties.

First, let us examine the dipole moment. This, although a vectorial property, is stationary; the dipole moment of a molecule *cannot* be affected by a symmetry operation. This immediately implies that the dipole moment vector *must* lie in each of the symmetry elements. In the case of the water molecule, or any other molecule in the point group C_{2v}, this is possible; a vector along the twofold axis lies in the axis, and also in each of the planes. Thus the symmetry argument has immediately determined the *direction* of the dipole moment. Only its *magnitude*, and which of its ends is positive and which negative, remain to be determined; these properties do not follow from symmetry considerations.

Examination of the applicable symmetry elements in other point groups immediately gives a large amount of information about dipole moments. Molecules belonging to point groups, where a center of symmetry applies, cannot have a dipole moment, since a vector cannot lie on a point. Similarly, molecules with more than one (noncoincident) rotational axis cannot have a dipole moment, since the vector cannot be coincident with two distinct axes. These considerations restrict the types of molecule which can have dipole moments to those belonging to just a few point groups. They must belong to point groups C_1, C_s, C_p, or C_{pv}. In the cases of C_p and C_{pv}, the direction of the dipole moment is determined; in C_s it is known to lie in the symmetry plane. Thus symmetry gives a great deal of qualitative information about dipole moments, or, in reverse, dipole moment measurements can give information about molecular symmetry and hence geometry.

Normal Vibrations. In a molecule consisting of n atoms, complete specification of the location in space of all atoms requires $3n$ coordinates, the so-called $3n$ degrees of freedom, that is, the three Cartesian coordinates for each atom. Of these $3n$ degrees of freedom, three are needed and are sufficient to determine the *position* of the molecule, that is, of its center of gravity, in space. Three more degrees are needed to define the *orientation* of the molecule, two angles to locate one axis (the principal axis), and one more angle to define the rotational position about this axis. These are the three so-called Euler angles (cf. Fig. 4.10). If the molecule is linear, and only then, different rotational positions about the principal axis are equivalent, and the last angle is

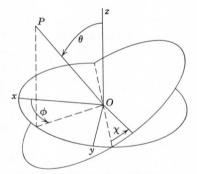

Fig. 4.10 The Euler angles. If OP is the principal axis of a molecule, its position is given by the angles θ and φ. χ measures the rotation about this axis.

meaningless and unnecessary. This leaves $3n - 6$ degrees of freedom (in a linear molecule $3n - 5$), which define the positions of the atoms relative to one another, and hence the bond distances and angles, and their changes, that is, the vibrations.

We have seen that it is possible to resolve a velocity (*translation*) or an angular momentum (*rotation*) into three components, corresponding to the three degrees of freedom they represent; similarly, it is possible to resolve an arbitrary *vibration* into as many components as the number of vibrational degrees of freedom ($3n - 6$, or, for the linear case, $3n - 5$). Such a resolution can be made in an infinite number of ways, just as there is an infinite number of ways a single vector can be resolved into three mutually orthogonal components.[2] But there was *one* preferred way to make the resolution of the velocity and angular momentum vectors—the one made above for water—by letting the components be along the axes of an

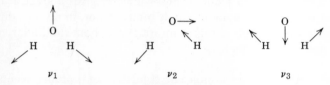

Fig. 4.11   The normal vibrations of water.

appropriately chosen coordinate system. Similarly, there is a preferred, usually unique way of making the resolution of an arbitrary vibration into components; this is such that the periodic motions of each of the atoms in any given component occur with precisely the same frequency. All future discussions of vibrational motion will be in terms of these components, which are called the *normal vibrations*.

Let us return to our example, the water molecule. In this triatomic molecule $3n - 6 = 9 - 6 = 3$, and we require three normal vibrations. These are most easily depicted by showing the directions of the motions of each atom. For water, without going into the ways in which these selections are made, they are shown in Fig. 4.11. Application of the symmetry operations appropriate to the water molecule to ν_1 and ν_3 shows that these remain unaffected under each of the operations, and hence transform as—or belong to—the species A_1. Under the operations I and $\sigma_v{}^{yz}$, ν_2 also remains unaffected, but under $C_2{}^z$ and $\sigma_v{}^{xz}$ each of the arrows drawn reverses direction; hence ν_2 transforms as B_2.

[2] For example, with the vector represented as starting from the origin of any particular coordinate system, we may rotate the coordinate system in any direction around the origin and obtain a new coordinate system with new x, y, and z properties.

It can be shown in general that each normal vibration transforms as one of the symmetry species of the point group of the molecule. This is of tremendous value in the vibrational analysis of any compound. In addition, the number of vibrations belonging to each symmetry species can be calculated readily from a knowledge of the location of the atoms with respect to the symmetry elements (cf. Chapter 5.)

The determination of the normal vibrations is a rather difficult problem, which fortunately concerns us little. In principle, it involves the solution of an equation of order $3n - 6$, that is, the solution for all values of x of an equation of the form:

$$x^{3n-6} + ax^{3n-5} + bx^{3n-4} + \cdots + kx^2 + lx + m = 0 \quad (4.3)$$

The coefficients $a, b, \ldots, k, l$ are complicated functions of the interatomic forces. This equation arises from expansion of a determinant of order $3n - 6$, which is equated to zero, and the elements of which depend on the interatomic forces. Obviously, the determination of the roots of such an equation, or, as frequently stated, the diagonalization of the determinant, is a formidable task. Even for the three vibrations of water this implies finding the three roots of a cubic equation, which involves considerable work. Parenthetically it may be added that all roots are, of necessity, real.

It can be shown, however, that the determinant may be factored; or in other words, that the equation (4.3) may be broken down into a product of a series of equations of much lower order by the simple use of symmetry. Each such product represents normal vibrations belonging to only a single symmetry species, and the *sum* of the orders of the separate equations is the order of the overall equation (4.3). Even in the relatively simple case of water this means that the cubic equation may be factored into a quadratic and a linear equation. This factorization, which depends exclusively on symmetry properties, is of extreme value in any practical work with normal vibrations.

4.2   DEGENERATE POINT GROUPS

The development given in the preceding section can be applied to the point groups C_1, C_s, C_i, C_2, C_{2v}, C_{2h}, D_2, and D_{2h}. All of these point groups are characterized by the *absence* of any symmetry element of order greater than two, that is, the absence of any axis C_p or S_p with $p > 2$. As soon as such an element arises, the problems of symmetry species become much more difficult.

Let us take as an example the molecule $XeOF_4$, one of the recently discovered compounds of xenon. Its structure is believed to be a tetragonal pyramid (Fig. 4.12) with the four F's in a square plane, and the Xe and O on the fourfold axis of this square, and hence it is thought to belong to the point group C_{4v}. Thus it may, and undoubtedly does, have a dipole moment which must be directed along the XeO axis.

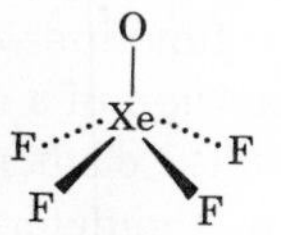

Fig. 4.12   Geometry of $XeOF_4$, point group C_{4v}.

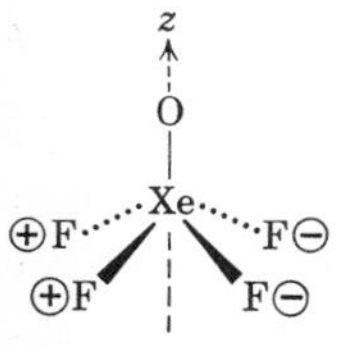

Fig. 4.13   Rotation of $XeOF_4$ about z. The circled pluses represent vectors pointed toward the reader; the circled minuses represent vectors away from the reader.

Treatment of a translation (velocity vector) along the z axis, which, by convention, is taken along the symmetry axis, produces no problems. This vector remains unaffected by any one of the eight symmetry operations, I, $2C_4$, C_2, $2\sigma_v$, $2\sigma_d$, and belongs to the totally symmetric species A_1, given in the first line of Table 4.2. Similarly, rotation about the z axis produces no new problems. Operation on the vectors shown in Fig. 4.13 shows that I, both C_4^z, and C_2^z, leave the motion unchanged. Reflection in any of the four planes, two passing through opposing F's, two bisecting opposite FXeF angles, reverses the direction of rotation, leading to the behavior pattern indicated in the second row of Table 4.2 and referred to as species A_2.

That this molecule behaves drastically differently from water becomes apparent, however, as soon as we examine the behavior of translation in the x direction, particularly under the operation C_4^z. Examination of Fig. 4.14 shows that $C_4'^z$, a counterclockwise rotation of the molecule around the z axis by 90°, transforms motion in the x direction into motion in the y direction. Thus $\mathbf{v}_x$ is neither

Table 4.2   Character Table for Point Group C_{4v}

C_{4v}	I	$2C_4$	C_2	$2\sigma_v$	$2\sigma_d$	
A_1	1	1	1	1	1	z
A_2	1	1	1	−1	−1	R_z
B_1	1	−1	1	1	−1	—
B_2	1	−1	1	−1	1	—
E	2	0	−2	0	0	$(x,y)(R_x,R_y)$

symmetric nor antisymmetric under this operation. A brief examination of Fig. 4.14 reveals that, at the same time, the operation $C_4'^z$ transforms motion in the y direction into motion in the $-x$ direction. Thus these two motions appear to be very closely connected.

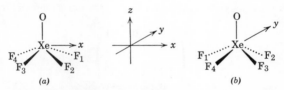

Fig. 4.14   Translation of $XeOF_4$ (a) in the x direction, and (b) as transformed by $C_4'^z$.

Transformation Matrices. We can discuss the above transformations in general terms by considering an arbitrary velocity vector $\mathbf{v}$ in any direction, having vector components $\mathbf{v}_x$ and $\mathbf{v}_y$ in the x and y directions respectively. On performing a counterclockwise rotation by 90° on the molecule placed in a fixed coordinate system, the new vector $\mathbf{v}'$ will have the components $\mathbf{v}_x'$ and $\mathbf{v}_y'$ given by

$$\mathbf{v}_x' = -\mathbf{v}_y$$
$$= 0\mathbf{v}_x - 1\mathbf{v}_y \tag{4.4}$$

and

$$\mathbf{v}_y' = \mathbf{v}_x$$
$$= 1\mathbf{v}_x + 0\mathbf{v}_y \tag{4.5}$$

Equations (4.4) and (4.5) tell us that the new $\mathbf{v}_x$, denoted by $\mathbf{v}_x{}'$, is equal to minus the old $\mathbf{v}_y$, and that the new $\mathbf{v}_y$, $\mathbf{v}_y{}'$, is equal to the old $\mathbf{v}_x$.

A special method of writing these two equations is possible:

$$\begin{pmatrix} \mathbf{v}_x{}' \\ \mathbf{v}_y{}' \end{pmatrix} = \begin{pmatrix} 0 & -1 \\ 1 & 0 \end{pmatrix} \begin{pmatrix} \mathbf{v}_x \\ \mathbf{v}_y \end{pmatrix} \tag{4.6}$$

This is called a matrix notation.

$$\begin{pmatrix} 0 & -1 \\ 1 & 0 \end{pmatrix}$$

is the *transformation matrix* which transforms the original set of velocities $\mathbf{v}_x$ and $\mathbf{v}_y$ into the new ones $\mathbf{v}_x{}'$ and $\mathbf{v}_y{}'$. To utilize this notation effectively, we must have some elementary knowledge of matrices and know the rules of matrix multiplication.

A matrix is a rectangular array of numbers, or symbols representing numbers; for example,

$$\begin{pmatrix} 2 & 0 & -1 \\ 9 & -1 & 4 \\ -2 & 3 & 0 \\ 0 & 5 & -1 \end{pmatrix}$$

The numbers are enclosed in parentheses, and since the array consists of four rows and three columns, this matrix is called a 4×3 matrix. A general symbolism has been developed and accepted for writing the elements of a matrix. The 4×3 matrix shown above is written as follows, where the value of a_{11} would be 2 and the value of a_{33} would be 0, and so on:

$$\begin{pmatrix} a_{11} & a_{12} & a_{13} \\ a_{21} & a_{22} & a_{23} \\ a_{31} & a_{32} & a_{33} \\ a_{41} & a_{42} & a_{43} \end{pmatrix}$$

The first digit in the subscript of the letter a indicates the row, and the second digit the column of the element. Thus a_{ij} represents the element in the ith row and the jth column. When the numbers of rows and columns are equal, the matrix is called a *square matrix*. The elements in a square matrix with $i = j$ are called the *diagonal elements* because they lie on a diagonal line from the upper left to the lower right corners of the square matrix, called the *principal diagonal*. In a 3×3 matrix the elements in the diagonal are $a_{11}a_{22}a_{33}$. A 3×1

matrix such as

$$\begin{pmatrix} x \\ y \\ z \end{pmatrix}$$

may be used to represent a vector in three-dimensional space.

We may now proceed to discuss the recipe for performing a *matrix multiplication*. This is achieved by multiplying the elements of the row by the elements of the column, each in turn. In the general form, the recipe for multiplication of a 2 × 2 matrix by a second 2 × 2 matrix is as follows:

$$\begin{pmatrix} a_{11} & a_{12} \\ a_{21} & a_{22} \end{pmatrix} \begin{pmatrix} b_{11} & b_{12} \\ b_{21} & b_{22} \end{pmatrix} = \begin{pmatrix} a_{11}b_{11} + a_{12}b_{21} & a_{11}b_{12} + a_{12}b_{22} \\ a_{21}b_{11} + a_{22}b_{21} & a_{21}b_{12} + a_{22}b_{22} \end{pmatrix}$$

In a real example using numbers,

$$\begin{pmatrix} 7 & 3 \\ 1 & 0 \end{pmatrix} \begin{pmatrix} 2 & 1 \\ 4 & 0 \end{pmatrix} = \begin{pmatrix} 7\cdot 2 + 3\cdot 4 & 7\cdot 1 + 3\cdot 0 \\ 1\cdot 2 + 0\cdot 4 & 1\cdot 1 + 0\cdot 0 \end{pmatrix} = \begin{pmatrix} 26 & 7 \\ 2 & 1 \end{pmatrix}$$

The first diagonal element of the product matrix, which is usually denoted as c_{11}, is the sum $a_{11}b_{11} + a_{12}b_{21}$; in the above example $c_{11} = 26$. If the matrices to be multiplied, instead of being 2 × 2 matrices, were $n \times m$ and $m \times l$ matrices, the product matrix would be $n \times l$ and the first element would be $c_{11} = a_{11}b_{11} + a_{12}b_{21} + a_{13}b_{31} + \cdots + a_{1m}b_{m1}$; and the element c_{ij} of the product matrix is the sum of all the products $a_{ik}b_{kj}$ for all values of the running index k from 1 to m.

$$c_{ij} = \sum_{k=1}^{m} a_{ik}b_{kj}$$

This formula shows that two matrices can be multiplied together only if the left one has as many columns as the right one has rows; such matrices are called conformable. That this condition is necessary is apparent since the index k runs simultaneously through all rows of the first and all columns of the second matrix. Matrix multiplication is not commutative, just as the multiplication of symmetry operations is not always commutative. Thus in the above multiplication of the two square matrices, if the order of multiplication were inverted to $\begin{pmatrix} 2 & 1 \\ 4 & 0 \end{pmatrix}\begin{pmatrix} 7 & 3 \\ 1 & 0 \end{pmatrix}$, the product $\begin{pmatrix} 15 & 6 \\ 28 & 12 \end{pmatrix}$ would be different from the one we calculated above. Matrix multiplication, however, like symmetry operation multiplication, is associative.

The transformations of vectors due to rotations, as well as many other transformations, are greatly facilitated by the use of matrices. To illustrate, we will now examine somewhat in detail the transformation of a pair of orthogonal (that is, perpendicular) vectors under a counterclockwise rotation by an arbitrary angle θ.[3] The two original vectors are shown in Fig. 4.15, as $\mathbf{r}_1$ and $\mathbf{r}_2$, and are assumed of equal length having components x_1, y_1, and x_2, y_2, respectively; that is, they are given by

$$\mathbf{r}_1 = x_1\mathbf{i} + y_1\mathbf{j}$$
$$\mathbf{r}_2 = x_2\mathbf{i} + y_2\mathbf{j} \tag{4.7}$$

The fact that they are orthogonal requires

$$\mathbf{r}_1 \cdot \mathbf{r}_2 = (x_1\mathbf{i} + y_1\mathbf{j}) \cdot (x_2\mathbf{i} + y_2\mathbf{j}) = 0$$
$$= x_1x_2\mathbf{i} \cdot \mathbf{i} + y_1y_2\mathbf{j} \cdot \mathbf{j} + (x_1y_2 + x_2y_1)\mathbf{i} \cdot \mathbf{j}$$
$$= x_1x_2 + y_1y_2 = 0$$

since $\mathbf{i} \cdot \mathbf{i} = \mathbf{j} \cdot \mathbf{j} = 1$ and $\mathbf{i} \cdot \mathbf{j} = 0$. Hence

$$\frac{x_1}{y_1} = \frac{-y_2}{x_2}$$

The equal length of the vectors requires:

$$x_1{}^2 + y_1{}^2 = x_2{}^2 + y_2{}^2$$

Squaring of the first and combination with the second of the last two equations gives:

$$\frac{x_1{}^2}{y_1{}^2} = \frac{y_2{}^2}{x_2{}^2}$$

$$\frac{x_2{}^2 + y_2{}^2 - y_1{}^2}{y_1{}^2} = \frac{y_2{}^2}{x_2{}^2}$$

$$\frac{x_2{}^2 + y_2{}^2}{y_1{}^2} - 1 = \frac{y_2{}^2}{x_2{}^2}$$

$$\frac{x_2{}^2 + y_2{}^2}{y_1{}^2} = 1 + \frac{y_2{}^2}{x_2{}^2} = \frac{x_2{}^2 + y_2{}^2}{x_2{}^2}$$

Hence $y_1{}^2 = x_2{}^2$, and either

$$x_1 = y_2 \quad \text{and} \quad y_1 = -x_2$$

or

$$x_1 = -y_2 \quad \text{and} \quad y_1 = x_2$$

The case shown in Fig. 4.15 is the first of these alternatives; the second refers to the vector $-\mathbf{r}_2$ similar to $\mathbf{r}_2$ but in opposite direction, which is also orthogonal

[3] The formulas here developed differ from those encountered in most mathematics books since it is customary to talk about the rotation of a *coordinate* system, holding the vector fixed, whereas we discuss the rotation of a vector in a fixed coordinate system. The two points of view are equivalent, except that a *counterclockwise* rotation of the vector in a fixed coordinate system corresponds to a *clockwise* rotation of the coordinate system with a fixed vector.

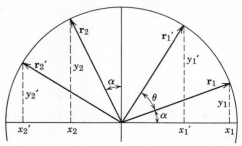

Fig. 4.15   The transformation of two orthogonal vectors, r_1 and r_2, upon rotation by θ.

to r_1.[4] From the relations between x_1, x_2, y_1, and y_2, we can rewrite both vectors in an alternate form:

$$r_1 = y_2 i - x_2 j; \qquad r_2 = -y_1 i + x_1 j \tag{4.8}$$

Upon rotation of r_1 and r_2 by θ, the two new vectors r_1' and r_2' result, which are given by:

$$r_1' = x_1' i + y_1' j; \qquad r_2' = x_2' i + y_2' j \tag{4.9}$$

To determine the relation of the rotated vectors to the original ones, we proceed as follows. x_1 and y_1 are the projections of r_1 on the x and y axes, given by:

$$x_1 = r_1 \cos \alpha, \qquad y_1 = r_1 \sin \alpha$$

where α is the angle between r_1 and the x axis.

Similarly,

$$
\begin{aligned}
x_1' &= r_1 \cos (\alpha + \theta) \\
&= r_1 \cos \alpha \cos \theta - r_1 \sin \alpha \sin \theta \\
&= x_1 \cos \theta - y_1 \sin \theta \\
y_1' &= r_1 \sin (\alpha + \theta) \\
&= x_1 \sin \theta + y_1 \cos \theta
\end{aligned}
$$

Substituting these relations into the expression for r_1' (eq. 4.9), we obtain:

$$
\begin{aligned}
r_1' &= (x_1 \cos \theta - y_1 \sin \theta) i + (x_1 \sin \theta + y_1 \cos \theta) j \\
&= \cos \theta (x_1 i + y_1 j) + \sin \theta (-y_1 i + x_1 j) \\
&= r_1 \cos \theta + r_2 \sin \theta
\end{aligned}
$$

[4] The use of the alternate vector $-r_2$ instead of r_2 in the counterclockwise rotation will lead to results which are identical with those obtained here, except that the off-diagonal elements of the transformation matrix have the signs exchanged.

Similarly, substituting the exactly analogous expressions for x_2' and y_2' into the expression for r_2' (eq. 4.9),

$$r_2' = (x_2 \cos \theta - y_2 \sin \theta)\mathbf{i} + (x_2 \sin \theta + y_2 \cos \theta)\mathbf{j}$$
$$= -\mathbf{r_1} \sin \theta + \mathbf{r_2} \cos \theta$$

or, in matrix notation,

$$\begin{pmatrix} \mathbf{r_1'} \\ \mathbf{r_2'} \end{pmatrix} = \begin{pmatrix} \cos \theta & \sin \theta \\ -\sin \theta & \cos \theta \end{pmatrix} \begin{pmatrix} \mathbf{r_1} \\ \mathbf{r_2} \end{pmatrix} \tag{4.10}$$

and using the rules for matrix multiplication,

$$\begin{pmatrix} \mathbf{r_1'} \\ \mathbf{r_2'} \end{pmatrix} = \begin{pmatrix} \mathbf{r_1} \cos \theta & +\mathbf{r_2} \sin \theta \\ -\mathbf{r_1} \sin \theta & +\mathbf{r_2} \cos \theta \end{pmatrix}$$

as above.

Suppose now we had a pair of vectors along the x and y axes. Application of the matrix transformation would give us:

$$\begin{pmatrix} \mathbf{r_1'} \\ \mathbf{r_2'} \end{pmatrix} = \begin{pmatrix} \cos 90° & \sin 90° \\ -\sin 90° & \cos 90° \end{pmatrix} \begin{pmatrix} \mathbf{r_1} \\ \mathbf{r_2} \end{pmatrix} \tag{4.11}$$

$$= \begin{pmatrix} 0 & 1 \\ -1 & 0 \end{pmatrix} \begin{pmatrix} \mathbf{r_1} \\ \mathbf{r_2} \end{pmatrix} \tag{4.12}$$

or $$\mathbf{r_1'} = \mathbf{r_2}, \qquad \mathbf{r_2'} = -\mathbf{r_1}$$

which tells us that the vector $\mathbf{r_1}$ is transformed by the rotation into the vector $\mathbf{r_2}$, and the vector $\mathbf{r_2}$ into a new one like $\mathbf{r_1}$, but in the opposite direction, which is, of course, painfully obvious. However, the general formula of eq. 4.10 will permit us to solve a problem involving rotation around any angle θ and any vector. The particular rotation of 90° from the x axis to the y axis was chosen as illustrative, because in many applications the degenerate p_y and p_x orbitals are transformed under a C_4 operation (Fig. 4.16); the transformation matrix

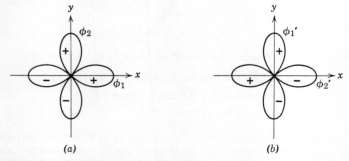

Fig. 4.16   Transformation of the original (a) p_x and p_y orbitals to the new orientation (b) on the C_4 operation; $\phi_1' = 0\phi_1 + 1\phi_2$; $\phi_2' = -1\phi_1 + 0\phi_2$.

is that shown in eq. (4.12). In the C_4' counterclockwise operation, the p_x orbital is transformed into the p_y orbital position (x into y), and simultaneously the p_y (degenerate with p_x) is transformed into the $-p_x$ orbital. We will continue the discussion in terms of the more general problem of transforming the vectors in the XeOF$_4$ molecule.

Returning now to the transformation eq. (4.6), we see that matrix multiplication gives:

$$\begin{pmatrix} \mathbf{v}_x' \\ \mathbf{v}_y' \end{pmatrix} = \begin{pmatrix} 0 & -1 \\ 1 & 0 \end{pmatrix}\begin{pmatrix} \mathbf{v}_x \\ \mathbf{v}_y \end{pmatrix} = \begin{pmatrix} 0\mathbf{v}_x - 1\mathbf{v}_y \\ 1\mathbf{v}_x + 0\mathbf{v}_y \end{pmatrix} = \begin{pmatrix} -\mathbf{v}_y \\ \mathbf{v}_x \end{pmatrix} \tag{4.13}$$

as shown above. In the future, instead of writing the specific transformation equations, we need only write the transformation matrix. Now it can be shown that the transformation of any symmetry operation can be characterized by the sum of the elements in the principal diagonal—the diagonal from upper left to lower right or the diagonal which is made up of terms with equal subscripts, a_{11}, a_{22}, etc.—of the transformation matrix. This summing of $a_{11} + a_{22} + \cdots$ has been called the *trace* (German *Spur*) of the matrix, and the actual numerical value of the trace is frequently called the *character* of the symmetry species or representation. In our present example the character is $0 + 0 = 0$.

Character Tables. Let us proceed to set up the transformation matrices for the x and y translation of our XeOF$_4$ molecule under the other operations. C_4^z, the rotation in the opposite sense (clockwise as Fig. 4.14 is drawn), transforms x into $-y$, y into x; consequently, the new x translation arises from the old y and the new y from the old $-x$, and hence we obtain the matrix $\begin{pmatrix} 0 & 1 \\ -1 & 0 \end{pmatrix}$ with a trace of 0. The identity transforms x into x, y into y, with a matrix $\begin{pmatrix} 1 & 0 \\ 0 & 1 \end{pmatrix}$ and a trace of 2. C_2 transforms x into $-x$, y into $-y$, with a matrix $\begin{pmatrix} -1 & 0 \\ 0 & -1 \end{pmatrix}$ and a trace of -2. The four σ are more complicated. Of the planes through the fluorine atoms, one transforms x into y and y into x, the other x into $-y$ and y into $-x$, with the matrices $\begin{pmatrix} 0 & 1 \\ 1 & 0 \end{pmatrix}$ and $\begin{pmatrix} 0 & -1 \\ -1 & 0 \end{pmatrix}$, respectively. Of the planes bisecting the FXeF angles, one transforms x into itself and y into $-y$, the other x into $-x$ and y into itself, corresponding to the matrices $\begin{pmatrix} 1 & 0 \\ 0 & -1 \end{pmatrix}$ and $\begin{pmatrix} -1 & 0 \\ 0 & 1 \end{pmatrix}$, respectively. Thus, for all four operations, the character is 0. These characters are to be found in line 5 of Table 4.2. We now see why this table is called a *character table*.

An interesting result arose in the last paragraph. We saw that the forward and reverse rotation C_4^z and $C_4'^z$ had different transformation matrices, but nevertheless identical characters; similarly

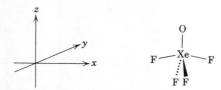

Fig. 4.17   An alternate assignment of coordinate axes in $XeOF_4$.

for the two planes σ_v and the two σ_d. This is a general result of sets of equivalent symmetry operations, called classes, and consequently these are not generally listed separately in a character table. As a matter of fact, the orientation of our molecule in the coordinate system, as shown in Fig. 4.14, is purely arbitrary. Although the z axis is fixed by the symmetry axis, we might as readily have put the x and y axes in one or the other of the OXeF planes, for example as in Fig. 4.17. If we did this, the transformation matrices for I, C_4^z, $C_4'^z$, and C_2^z would be unaffected, but the four σ and their transformation matrices would be permuted among each other; still the characters are unaffected. Actually, it can be shown that we can arbitrarily place the x and y axes at any desired direction, as long as they are perpendicular (orthogonal) to each other and to the z axis. Again, the transformation matrices change depending on the axes chosen, but the traces are invariant.

This equivalence is a general property of any two vector or tensor quantities that transform under symmetry operations into one another, and they are called *degenerate*. The symmetry species to which they belong are also called degenerate, and designated by the symbol E as long as there are just pairs of degenerate quantities and the character of I is 2. We will meet three- and higher-fold degenerate species later on.

Let us momentarily return to the previous example, the water molecule. Here, each quantity transformed under symmetry operations into itself, so that each transformation matrix was the 1×1 matrix (1) or (-1) with character 1 or -1. Thus the simpler case of the point group C_{2v} is included in our present treatment, only the lack of degeneracy has made it so much simpler. Similarly, the transformation matrices of species A_1 and A_2 of C_{4v} are 1×1 matrices, $(+1)$ or (-1), with traces of $+1$ or -1.

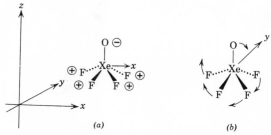

Fig. 4.18   The rotation of $XeOF_4$ about (a) x and (b) y.

We shall now proceed to the rotations about the x and y axes for $XeOF_4$. Figure 4.18 shows these rotations. The reader may verify that the transformation matrices for the operations indicated are

$$I\begin{pmatrix} 1 & 0 \\ 0 & 1 \end{pmatrix}; \quad C_4^z\begin{pmatrix} 0 & 1 \\ -1 & 0 \end{pmatrix}; \quad C_4'^z\begin{pmatrix} 0 & -1 \\ 1 & 0 \end{pmatrix};$$

$$C_2^z\begin{pmatrix} -1 & 0 \\ 0 & -1 \end{pmatrix}; \quad \sigma_v^{yz}\begin{pmatrix} -1 & 0 \\ 0 & 1 \end{pmatrix}; \quad \sigma_v^{xz}\begin{pmatrix} 1 & 0 \\ 0 & -1 \end{pmatrix};$$

$$\sigma_d\begin{pmatrix} -1 & 0 \\ 0 & 1 \end{pmatrix}; \quad \text{and} \quad \begin{pmatrix} 1 & 0 \\ 0 & -1 \end{pmatrix}$$

The characters then are 2, 0, 0, -2, 0, 0, 0, and 0, respectively—the same as for the translations. The rotations are equivalent and degenerate and belong to the species E. It may thus be concluded from symmetry considerations that the moments of inertia about the x and y axes are identical, and consequently the energies of rotational motion are equal (at equal angular velocity), so that the rotations are degenerate in the sense of having equal energies. It can also be shown that the moments of inertia about the alternately defined axes of Fig. 4.17 are the same as those about the axes of Fig. 4.18, and the same about any axes through the center of gravity and perpendicular to the z axis. This, then, is the essential meaning of the degeneracy—that the quantities occur in completely equivalent and indistinguishable pairs.

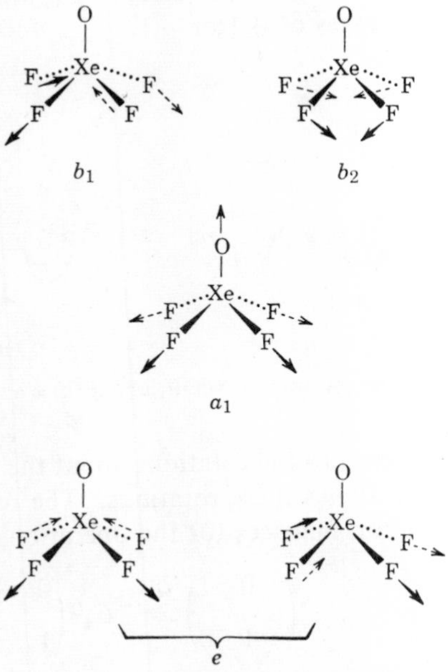

Fig. 4.19   Vibrations of $XeOF_4$.

We have not, however, yet exhausted the character table for point group C_{4v}. While translations and rotations belong to species A_1, A_2, and E, we can find vibrations and wave functions belonging to the additional species B_1 and B_2, for which the characters are included in Table 4.2.

$XeOF_4$ has 6 atoms and hence $18 - 6 = 12$ normal vibrations. Although we will not attempt to draw all of these, Fig. 4.19 gives examples of vibrations belonging to all symmetry species, with the exception of A_2, to which no normal vibrations belong. It is left to the reader to verify their symmetry properties. One could equally well draw wave functions (orbitals) for each of the symmetry species, but the diagrams would tend to become very complicated.

Sec. 4.2   *Degenerate Point Groups*   83

Character Table for Benzene, D_{6h}. In the example of $XeOF_4$ (C_{4v}), all elements of the transformation matrices were either 0 or ± 1, and any symmetry operation transformed any quantity into itself, its negative, or the positive or negative of its degenerate orthogonal equivalent. This is a characteristic of a fourfold axis and does not hold for any axis of order other than 2 or 4. Let us next examine benzene, which belongs to the point group D_{6h}. The sixfold axis is the z axis, and motion in its direction, or rotation about it, is readily

Table 4.3   Symmetry Species and Characters for the Point Group D_{6h}

D_{6h}	I	$2C_6(z)$	$2C_6{}^2 \equiv 2C_3$	$C_6{}^3 \equiv C_2{}''$	$3C_2$	$3C_2{}'$	σ_h	$3\sigma_v$	$3\sigma_d$	$2S_6$	$2S_3$	$S_6{}^3 \equiv S_2 \equiv i$
A_{1g}	+1	+1	+1	+1	+1	+1	+1	+1	+1	+1	+1	+1
A_{1u}	+1	+1	+1	+1	+1	+1	−1	−1	−1	−1	−1	−1
A_{2g}	+1	+1	+1	+1	−1	−1	+1	−1	−1	+1	+1	+1
A_{2u}	+1	+1	+1	+1	−1	−1	−1	+1	+1	−1	−1	−1
B_{1g}	+1	−1	+1	−1	+1	−1	−1	−1	+1	+1	−1	+1
B_{1u}	+1	−1	+1	−1	+1	−1	+1	+1	−1	−1	+1	−1
B_{2g}	+1	−1	+1	−1	−1	+1	−1	+1	−1	+1	−1	+1
B_{2u}	+1	−1	+1	−1	−1	+1	+1	−1	+1	−1	+1	−1
E_{1g}	+2	+1	−1	−2	0	0	−2	0	0	−1	+1	+2
E_{1u}	+2	+1	−1	−2	0	0	+2	0	0	+1	−1	−2
E_{2g}	+2	−1	−1	+2	0	0	+2	0	0	−1	−1	+2
E_{2u}	+2	−1	−1	+2	0	0	−2	0	0	+1	+1	−2

shown to belong to species A_{2u} or A_{2g}, respectively, with 1×1 transformation matrices and the characters shown in Table 4.3. But the interesting problems arise out of operation by C_6 $C_6{}'$, $C_6{}^2$, and $C_6{}'^2$ on the motion in the x and y direction. For example, $C_6{}'$ transforms a vector in the x direction into one in a direction 60° from x between x and y, and thus (eq. 4.10):

$$\mathbf{v}_x{}' = \cos 60\, v_x + \sin 60\, v_y$$
$$\mathbf{v}_y{}' = -\sin 60\, v_x + \cos 60\, v_y$$

and the transformation matrix is

$$\begin{pmatrix} \cos 60 & \sin 60 \\ -\sin 60 & \cos 60 \end{pmatrix}$$

with a trace of $2\cos 60$ and therefore a character of 1. Similarly, the transformation matrices for the x and y translation under the operations C_6 (clockwise), $C_6'^2$ (that is, rotation by $120°$ counterclockwise) and C_6^2 (clockwise) are:

$$\begin{pmatrix} \cos 60 & -\sin 60 \\ \sin 60 & \cos 60 \end{pmatrix}, \quad \begin{pmatrix} \cos 120 & \sin 120 \\ -\sin 120 & \cos 120 \end{pmatrix}, \quad \begin{pmatrix} \cos 120 & -\sin 120 \\ \sin 120 & \cos 120 \end{pmatrix}$$

respectively, with traces of $2\cos 60 = 1$, $2\cos 120 = -1$, and $2\cos 120 = -1$. These transformation matrices then show that, under these operations, the translations transform not into themselves nor into their degenerate equivalent but into a linear combination of the two. The reader can verify that this is equally true for some—but not all—of the planes σ_v. The transformation matrices, if constructed, will give the characters shown under species E_{1u} of Table 4.3. The u arises by application of the operation i which transforms $\mathbf{v}_x$ into $-\mathbf{v}_x$, $\mathbf{v}_y$ into $-\mathbf{v}_y$ by the transformation matrix $\begin{pmatrix} -1 & 0 \\ 0 & -1 \end{pmatrix}$ yielding the trace -2. It may be added that the transformation matrix for the operation i is always a diagonal matrix—one in which all terms not on the principal diagonal are zero—and that the diagonal elements must be all $+1$ or all -1, depending on whether the species has g or u character.

Application of the same procedures to the rotations about the x and y axes gives transformation matrices which have the characters given under the symmetry species E_{1g} of D_{6h} (Table 4.3). No new principles are involved. It seems hardly worthwhile to reproduce all the thirty normal vibrations of benzene, nor even to give examples of one for each of the twelve symmetry species. However, consideration of some of the molecular orbitals may be very instructive.

First, let us examine if symmetry places any restrictions on the symmetry species to which the π molecular orbitals of benzene may belong. These orbitals are formed from the p_z atomic orbitals of the six carbon atoms. Since each of these atomic orbitals is antisymmetric relative to the molecular plane (σ_h), the molecular orbitals must belong to symmetry species which have a character of -1 for σ_h, except for doubly degenerate species, which must have a character of -2. This restricts the π molecular orbitals to the species a_{1u}, a_{2u}, b_{1g}, b_{2g}, e_{1g}, and e_{2u} (cf. Table 4.3). The lowest π-electron molecular orbital is shown in Fig. 4.20 and given by

$$\psi_1 = \frac{1}{\sqrt{6}}(\phi_1 + \phi_2 + \phi_3 + \phi_4 + \phi_5 + \phi_6)$$

where the ϕ's are the atomic $2p\pi$ orbitals of the six carbon atoms. It is readily seen that all rotations about the z axis, C_6, C_6^2, and so on, transform this orbital into itself, as do all the vertical symmetry planes and I. The horizontal plane σ_h, on the other hand, the center i, all the C_2 in the molecular plane and the rotation reflection axes coincident with C_6 (S_6, and so on) transform this orbital into its negative, since positive and negative areas are exchanged. Thus each operation transforms the orbital into itself or its negative, and the characters are $+1$ or -1 and

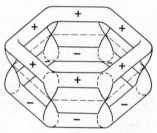

Fig. 4.20 ψ_1 of benzene, $(1/\sqrt{6})$· $(\phi_1 + \phi_2 + \phi_3 + \phi_4 + \phi_5 + \phi_6)$.

correspond to the species a_{2u}. Thus this orbital is nondegenerate, as is well known, since it is the only orbital having energy 2β in simple Hückel MO theory.

The next higher two orbitals, however, arise in the calculation as a degenerate pair of energy β; they have the form shown in Fig. 4.21a, and are given by

$$\psi_2 = \frac{1}{2}(\phi_2 + \phi_3 - \phi_5 - \phi_6)$$

$$\psi_3 = \frac{1}{\sqrt{12}}(2\phi_1 + \phi_2 - \phi_3 - 2\phi_4 - \phi_5 + \phi_6)$$

(4.14)

Application of C_6 to these functions produces the new functions, shown in Fig. 4.21b:

$$\psi_2' = \frac{1}{2}(\phi_3 + \phi_4 - \phi_6 - \phi_1)$$

$$\psi_3' = \frac{1}{\sqrt{12}}(\phi_1 + 2\phi_2 + \phi_3 - \phi_4 - 2\phi_5 - \phi_6)$$

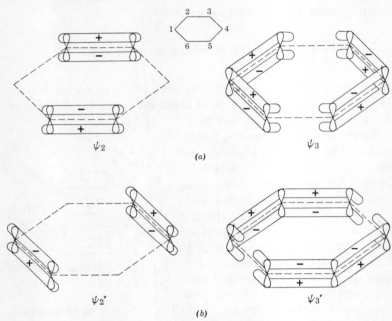

Fig. 4.21   (a) The degenerate wave functions ψ_2 and ψ_3 in benzene, and (b) the rotation of ψ_2 and ψ_3 about C_6 by $\pi/3$.

These are linear combinations of the original functions:

$$\psi_2{}' = \psi_2 \cos 60° - \psi_3 \sin 60° = \frac{1}{2}\,\psi_2 - \frac{\sqrt{3}}{2}\,\psi_3$$

$$\psi_3{}' = \psi_2 \sin 60° + \psi_3 \cos 60° = \frac{\sqrt{3}}{2}\,\psi_2 + \frac{1}{2}\,\psi_3 \tag{4.15}$$

as can be verified by substituting the expression in eq. (4.14) into eq. (4.15). Thus the transformation matrix is $\begin{pmatrix} 1/2 & -\sqrt{3}/2 \\ \sqrt{3}/2 & 1/2 \end{pmatrix}$ and the trace $+1$. Table 4.4 gives the matrix elements of the application of all applicable symmetry elements for this degenerate pair of orbitals ψ_2 and ψ_3, and the resulting traces. Comparison with the character table shows the pair to belong to the species E_{1g}.

Character Tables for Linear Molecules.   We have thus seen the various complications which can arise from a multiple symmetry axis. In linear molecules (diatomics like H_2 or HCl, polyatomics like CO_2, acetylene), the lengthwise axis is infinite-fold and produces new problems. The first of these is that there is an infinite number of symmetry elements: The lengthwise axis is simultaneously C_p

Table 4.4 Elements for the Transformation of ψ_2 and ψ_3 of Benzene under the Various Symmetry Operations of the Point Group D_{6h}

	a_{11}	a_{12}	a_{21}	a_{22}	
I	1	0	0	1	
C_6	1/2	$-\sqrt{3}/2$	$\sqrt{3}/2$	1/2	
C_6'	1/2	$\sqrt{3}/2$	$-\sqrt{3}/2$	1/2	
$C_3 = C_6^2$	$-1/2$	$\sqrt{3}/2$	$-\sqrt{3}/2$	$-1/2$	
$C_3' = (C_6')^2$	$-1/2$	$-\sqrt{3}/2$	$\sqrt{3}/2$	$-1/2$	
$C_2'' = C_6^3$	-1	0	0	-1	
C_2	$+1$	0	0	-1	through atoms 1 and 4
C_2	$-1/2$	$-\sqrt{3}/2$	$-\sqrt{3}/2$	1/2	through atoms 2 and 5
C_2	$-1/2$	$\sqrt{3}/2$	$\sqrt{3}/2$	1/2	through atoms 3 and 6
C_2'	1/2	$-\sqrt{3}/2$	$-\sqrt{3}/2$	$-1/2$	through 1,2 and 4,5 bonds
C_2'	-1	0	0	$+1$	through 2,3 and 5,6 bonds
C_2'	1/2	$\sqrt{3}/2$	$\sqrt{3}/2$	$-1/2$	through 3,4 and 1,6 bonds
σ_h	-1	0	0	-1	
σ_v	-1	0	0	$+1$	through atoms 1 and 4
σ_v	1/2	$\sqrt{3}/2$	$\sqrt{3}/2$	$-1/2$	through atoms 2 and 5
σ_v	1/2	$-\sqrt{3}/2$	$-\sqrt{3}/2$	$-1/2$	through atoms 3 and 6
σ_d	$-1/2$	$\sqrt{3}/2$	$\sqrt{3}/2$	1/2	through 1,2 and 4,5 bonds
σ_d	$+1$	0	0	-1	through 2,3 and 5,6 bonds
σ_d	$-1/2$	$-\sqrt{3}/2$	$-\sqrt{3}/2$	1/2	through 3,4 and 1,6 bonds
S_6	$-1/2$	$\sqrt{3}/2$	$-\sqrt{3}/2$	$-1/2$	
S_6'	$-1/2$	$-\sqrt{3}/2$	$\sqrt{3}/2$	$-1/2$	
S_3	1/2	$-\sqrt{3}/2$	$\sqrt{3}/2$	1/2	
S_3'	1/2	$\sqrt{3}/2$	$-\sqrt{3}/2$	1/2	
i	1	0	0	1	

for *any* value of p, C_p' for any $p > 2$, and $(C_p)^q$ and $(C_p')^q$ as long as $q < p/2$. In addition, any plane containing the z axis is a symmetry plane, and there is an infinite number of these. Obviously, not all the elements or their characters can be written. A second difficulty arises from the notation customarily used for the symmetry species, which is completely different from that previously introduced.

Take as an example the molecule OCS, of point group $C_{\infty v}$. A translation along the C_∞ axis (the z axis) is obviously totally symmetric; it transforms into itself under all operations C_p and σ_v. But translations at right angles to this axis may be expected to be degenerate—our experience in preceding sections has shown translations normal to a multiple axis to be degenerate. Depending on the angle of rotation, then, they are rotated into some linear combination of each other. The most convenient way of writing the operation of rotation about an infinite-fold axis is to write it as a rotation about an arbitrary angle φ. Then the transformation matrices for counterclockwise and clockwise rotations, respectively, become

$$\begin{pmatrix} \cos\varphi & \sin\varphi \\ -\sin\varphi & \cos\varphi \end{pmatrix} \quad \text{and} \quad \begin{pmatrix} \cos\varphi & -\sin\varphi \\ \sin\varphi & \cos\varphi \end{pmatrix}$$

both with traces of $2\cos\varphi$. If rotation by an arbitrary angle φ is a symmetry operation, so is, of course, rotation by $2\varphi, 3\varphi, \ldots,$ $n\varphi$. For these operations, the traces obviously become $2\cos 2\varphi$, $2\cos 3\varphi, \ldots, 2\cos n\varphi$. Reflection on the xz plane transforms $\mathbf{v}_x$ into itself and $\mathbf{v}_y$ into $-\mathbf{v}_y$, and hence the trace for this plane is 0. Since all σ_v are equivalent, their traces are 0, even though the transformation matrices for the different planes differ.

Similar application of the various operations to the rotations about the x and y axes gives the same set of characters, and hence the same symmetry species. Unfortunately, rotation about the z axis is not a physically significant process for a molecule; however, if we consider a piece of pipe with a C_∞ axis, we can clearly observe a rotational motion about the lengthwise axis, and note that it remains unchanged by the symmetry operation C_∞ but is reversed by any σ_v. In the notation we have used so far, we would have called the totally symmetric species A_1, that of the rotational motion about z A_2, and the sideways translations or rotations E (or better, E_1, since there are other E's). The actual notation used, however, is borrowed from quantum mechanics, which in turn adapted it from early notation of atomic spectroscopy. A wave function belonging to one of the species we would like to call A has no angular momentum about the z axis. In atomic cases a function without angular momentum is called s (for one electron) or S (for many). Similarly, in the linear molecule we use the Greek equivalent, σ or Σ. The first type of degenerate function in the atomic case has one unit of angular momentum, and is called p or P; in the molecular case our degenerate functions are π or Π. Thus, translation along z and rotational

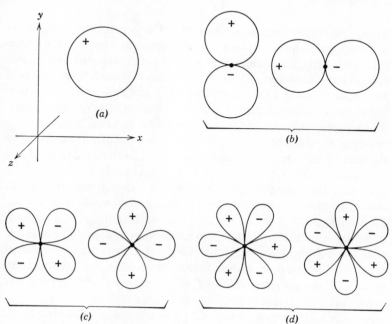

Fig. 4.22 Symmetry species of wave functions of linear molecules (*a*) σ, (*b*) π, (*c*) δ, and (*d*) ϕ.

motion about it have become of species Σ; however, we have seen that these two functions belong to different species, which are distinguished as Σ^+ and Σ^-, depending on the behavior (symmetric or antisymmetric) with respect to σ_v. The degenerate species encountered so far is Π (and no $+$, $-$ distinction is necessary).

Additional species are necessary in $C_{\infty v}$. Figure 4.22 shows wave functions of a linear molecule. It is apparent that functions with more and more nodal planes and more and more lobes can be written, corresponding to the functions s, p, d, f, g, ... of the atomic case. For the linear molecule, these are called σ, π, δ, φ, γ, ..., or

Σ, Π, Δ, Φ, Γ, . . . , and all but the first (σ,Σ) are doubly degenerate, as examination of Fig. 4.22 shows.

The addition of a center of symmetry i and a horizontal plane σ_h to form point group $D_{\infty h}$ of CO_2 produces no really new problem. Each symmetry species of $C_{\infty v}$ simply splits up into two, one g, the other u.

Triply and Higher Degenerate Species.   The point groups considered up to this point have involved no more than one axis of order higher than two. The very common classes of octahedral and tetrahedral compounds, however, all have four threefold and three fourfold axes (the fourfold axes are rotational in the octahedral, rotation-reflection for tetrahedral compounds). In these types of compounds, for example, CCl_4 and SF_6, some new problems arise. Consider, for example, the translation (velocity vector) in the x direction, assuming the coordinates assigned as shown in Figs. 4.23a and 4.24. Rotation about one of the threefold axes (for example, the C—Cl_1 bond) in the clockwise and counterclockwise direction, respectively, transforms this translation into translations into the z and y directions as shown, for example, in Figs. 4.23b and c. Thus the three translations appear to be degenerate. Similarly, the rotations about x, y, and z axes are transformed into each other by the S_4 or C_4 operations. Thus it appears that these point groups permit threefold degenerate symmetry species. These are referred to as T, or sometimes F, with appropriate subscripts. The corresponding character tables are given in Appendix 1. Figures 4.25 and 4.26 give vibrations belonging to the various symmetry species; for SF_6 there are no vibrations transforming like A_{1u}, A_{2g}, A_{2u}, E_u, or T_{1g}; for CCl_4 none transform like A_2 or T_1, and hence these species are lacking from the figures.

The point groups I and I_h are rarely encountered in molecules. The only molecule known to belong to I_h is $B_{12}H_{12}{}^{2-}$. In these point groups, in addition to the species so far encountered, four- and five-fold degenerate species also occur. The appropriate character tables are given in Appendix 1.

The last point group of interest is K_h, the completely symmetric group of spherical symmetry (K for Kugel group). This is the point group applicable to the free atom. It permits I, i, an infinite number of C_p and S_p (for each p from one to infinity), and an infinite number of σ. There also is an infinite number (1 to ∞) of symmetry species

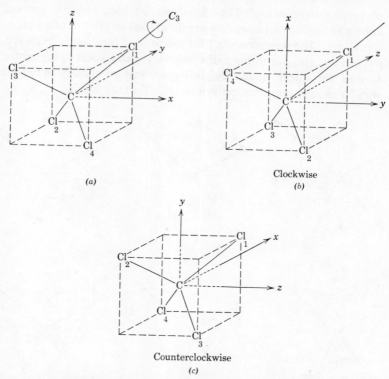

Fig. 4.23   The degeneracy of the Cartesian coordinates in tetrahedral molecules.

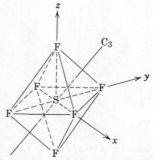

Fig. 4.24   The degeneracy of the Cartesian coordinates in octahedral molecules.

of varying degeneracy. The characters, at least in part, are well known to the chemist since they are the classes of orbitals of an atom, s, p, d, etc. The degeneracy is 1 for s, 3 for p, 5 for d, etc.; and s is g, p is u, d is g, etc., according to the behavior with respect to the center. Just as for linear molecules, this classification focuses more on angular momentum than on symmetry. Further species, s_u, p_g, d_u,

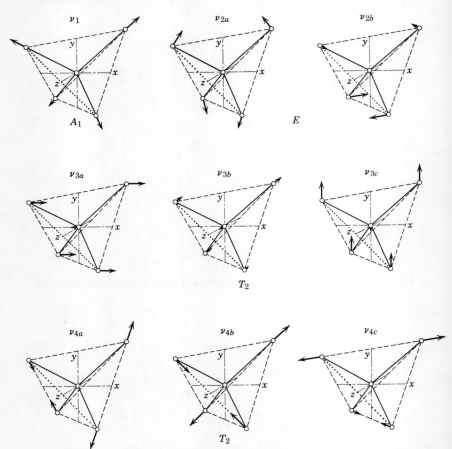

Fig. 4.25   Normal vibrations of a tetrahedral XY_4 molecule. The three twofold axes (dashed lines) are chosen as x, y, and z axes. (Reprinted by permission from G. Herzberg, *Molecular Spectra and Molecular Structure*, Vol. 2, *Infrared and Raman Spectra of Polyatomic Molecules*, D. Van Nostrand Co., Princeton, N.J., 1945.)

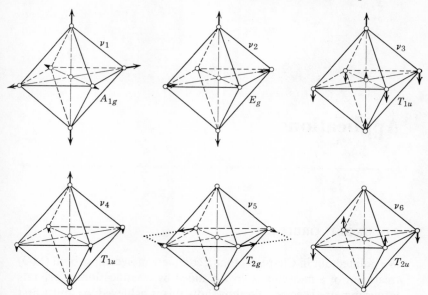

Fig. 4.26   Normal vibrations of an octahedral XY_6 molecule (point group O_h). Only one component of each degenerate vibration is shown. (Reprinted by permission from G. Herzberg, *Molecular Spectra and Molecular Structure*, Vol. 2, *Infrared and Raman Spectra of Polyatomic Molecules*, D. Van Nostrand Co., Princeton, N.J., 1945.)

etc., should be possible, but they are of little interest. For example, the rotational motion of a sphere about any axis belongs to p_g. The character table for this point group is of little interest and is not given.

REFERENCES

Gerhard Herzberg, *Molecular Spectra and Molecular Structure*, Vol. 2, *Infrared and Raman Spectra of Polyatomic Molecules*, D. Van Nostrand Co., Princeton, N.J., 1945, Chapter 2, Sections 2 and 3(d).

E. B. Wilson, Jr., J. C. Decius, and P. C. Cross, *Molecular Vibrations*, McGraw-Hill Book Co., New York, 1955, Chapter 5.

$$5$$

Applications

....

5.1 GROUP ORBITALS

We pointed out in Chapter 4 that a complete molecular orbital (MO) treatment of a molecule is greatly aided by classification of orbitals into symmetry species. To show how this is achieved, we must first sketch how the treatment would proceed without this simplification. Assume that we want to treat a molecule consisting of n atoms; for example, the SF_6 molecule with $n = 7$. We need to consider m atomic orbitals, the valence shell orbitals consisting of one 2s and three 2p orbitals for each F atom, and the 3s and three 3p's of S, for a total of $m = 28$, called the basis set. The individual one-electron MO's would then be linear combinations of all these m (in SF_6, 28) orbitals. Determination of the energies of the m resulting MO's, and of the m^2 (that is, $28 \times 28 = 784!$) coefficients in the linear combination requires the solution of an mth-order equation for the m roots (all of which are real), or alternately, diagonalization of an $m \times m$ determinant. For molecules of any appreciable size, this becomes a formidable task, and only high-speed computers are capable of accomplishing it. Even for a basis set of no more than six orbitals, hand calculation becomes prohibitive or nearly so.

In this situation classification of the basis orbitals in terms of symmetry species becomes of major importance. This is true because the matrix elements formed from orbitals of different symmetry types vanish. The matrix element is an integral of the form

$$\int \phi_1 H \phi_2 \, d\tau$$

where H is the Hamiltonian operator, which is always totally symmetric, and it can readily be shown that this integral vanishes if ϕ_1 and ϕ_2 belong to different symmetry species. Provided all atomic orbitals forming the basis set transform as one of the species of the molecule, the determinant can be factored into a product of several separate determinants. This condition is fulfilled, for example, in CO, and using a basis set of 8 orbitals, 2s and three 2p from either atom, the 8×8 determinant factors into a 4×4 determinant and two degenerate (and hence identical) 2×2 determinants. The 4×4 determinant involves the 2s and 2pσ orbitals of species Σ^+ of each atom, each 2×2 involves one 2pπ orbital of species Π from each atom.

Unfortunately, however, this ideal situation is almost never realized. With the exception of heteronuclear diatomics, and other linear molecules of point group $C_{\infty v}$, it occurs only for unsymmetric molecules of point group C_1, where it is immaterial since only one species A exists, and for planar molecules of point group C_s, where only two species exist; in other words, only if *all* atoms lie on *each* symmetry element.

In the very simple molecule water, the oxygen atom lies on all symmetry elements, and hence its orbitals transform as the symmetry species of C_{2v} of water namely A_1 (1s, 2s, 2p$_z$), B_1 (2p$_x$), and B_2 (2p$_y$). However, the 1s orbitals of each individual H atom fail to transform as any of the four symmetry species of C_{2v}. To overcome this difficulty, we combine the orbitals of the H atoms into so-called group or symmetry orbitals, by forming linear combinations:

$$\phi(a_1) = \frac{1}{\sqrt{2}} (1s_A + 1s_B)$$

$$\phi(b_2) = \frac{1}{\sqrt{2}} (1s_A - 1s_B) \tag{5.1}$$

which are shown schematically in Fig. 5.1. The formulation of eq. (5.1) is approximate since it is normalized only if the atomic orbitals $1s_A$ and $1s_B$ of H atoms A and B, respectively, do not overlap, that is, if $S_{AB} = \int 1s_A 1s_B \, d\tau = 0$. Under this assumption, also, the energies of $\phi(a_1)$ and $\phi(b_2)$ are equal and the same as those of either $1s_A$ or $1s_B$. If this approximation is not made, each orbital is to be multiplied by an additional normalization factor, $1/\sqrt{1 \pm S_{AB}}$, and their energies are the solution of a quadratic equation, but their forms are unchanged.

The symmetry species of these symmetry orbitals are shown in Table 5.1, and listed together with the appropriate O orbitals. Now

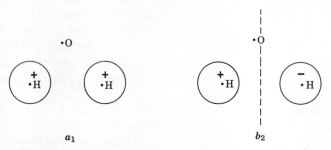

$$a_1 \qquad\qquad\qquad b_2$$

Fig. 5.1   The group orbitals formed from the hydrogen atoms of water.

the total 6×6 determinant of H_2O factors into a 3×3 for A_1, a 1×1 for B_1, and a 2×2 for B_2. In other words, the problem reduces from the solution of a sixth-order equation to that of the solution of one third, one second, and one first-order equation. The

**Table 5.1   Symmetry
Orbitals in H_2O**

	O	H_2
A_1	s, p_z	$1s_A + 1s_B$
A_2		
B_1	p_x	
B_2	p_y	$1s_A - 1s_B$

electronic structure may be described in the following way. The H_2 group orbital of species a_1 (Fig. 5.1) combines with the s and p_z orbitals of oxygen to form the a_1 molecular orbital shown in Fig. 5.2. The H_2 group orbital, b_2 of Fig. 5.1, combines with the p_y orbital of O to form the second bonding molecular orbital of water of species b_2 shown in Fig. 5.2. In addition to these bonding orbitals, there will be two (lone-pair) electrons in each of the two nonbonding orbitals, one of which is the p_x orbital of O and the other another a_1 orbital formed almost exclusively from the s and p_z orbitals of oxygen.

Naturally, the simplification obtained by factoring the determinant carries a price tag. The integrals that are the elements of the factored determinants are not the same as those of the original 6×6 determinants; they must be obtained by appropriate linear combinations of the elements of the original determinant. This, however, is small price to pay for the simplification obtained.

a_1 b_2

Fig. 5.2 The molecular orbitals of water formed from the oxygen s, p_z, and p_y atomic orbitals and the group orbitals of the hydrogen atoms.

The problem is analogous for more complex molecules. Consider, for example, the $PtCl_4^{2-}$ ion which belongs to point group D_{4h}. The steps in developing a qualitative description of the bonding in this ion are:

1. Assignment of the platinum atomic orbitals to the symmetry species in point group D_{4h}.

2. Formation of group orbitals from the s and p chlorine atomic orbitals and assignment of these orbitals to the appropriate symmetry species of D_{4h}.

3. Hybridization of the group orbitals of the same symmetry; although this step may be eliminated, it leads to a simpler and more readily visualized result.

4. Combination of the Pt atomic and Cl_4 group orbitals of similar symmetry to form the molecular orbitals.

Figure 5.3 shows five 5d, one 6s, and three 6p platinum orbitals; the five degenerate d and three degenerate p orbitals of the free metal atom have been assigned to the symmetry species appropriate to the square planar configuration (D_{4h}). The symmetry species of each orbital was obtained by the method illustrated in the following examples.

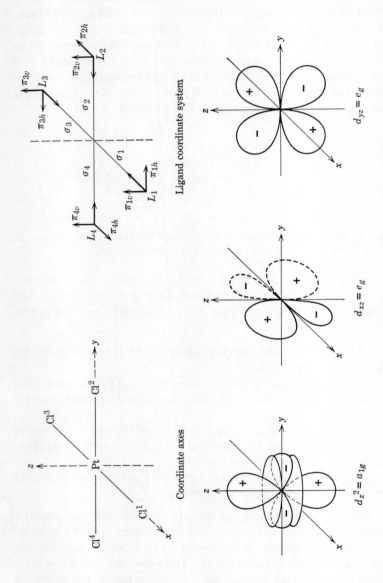

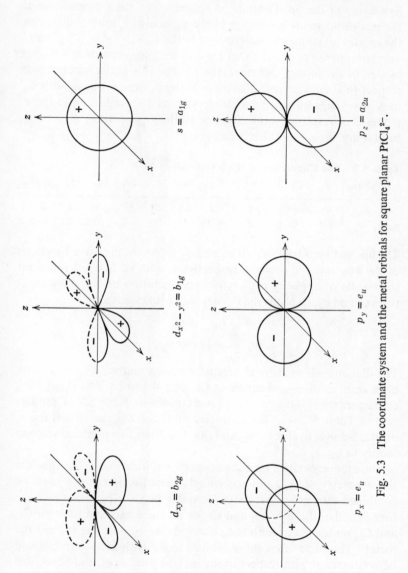

Fig. 5.3 The coordinate system and the metal orbitals for square planar PtCl₄²⁻.

Symmetry of the $6p_z$ **Orbital.**   Application of the C_4 operation to the p_z orbital yields no change in the signs of the lobes; therefore, the orbital is symmetric with respect to the C_4 axis. Applying the C_2 operation to the p_z about either the x or the y axis results in a change of sign of the lobes; therefore, the p_z orbital is antisymmetric with respect to these C_2 axes. Rotation of the p_z orbital about both $C_2{'}$ axes also yields an antisymmetric result, as does reflection in the σ_h plane. The results of application of all the D_{4h} operations on the p_z orbital are shown in Table 5.2.

Table 5.2   The Characters of the p Orbitals of Pt

D_{4h}		I	$2C_4$	$C_2{''} \equiv C_4{}^2$	$2C_2$	$2C_2{'}$	σ_h	$2\sigma_v$	$2\sigma_d$	$2S_4$	$S_2 \equiv i$
p_z	a_{2u}	+1	+1	+1	−1	−1	−1	+1	+1	−1	−1
p_x, p_y	e_u	+2	0	−2	0	0	+2	0	0	0	−2

The $6p_x$ **and** $6p_y$ **Orbitals.**   If ψ_1 and ψ_2 represent the wave functions of the $6p_x$ and $6p_y$ orbitals, respectively, and $\psi_1{'}$ and $\psi_2{'}$ represent the results of applying a C_4 symmetry operation on the wave functions ψ_1 and ψ_2, respectively, the new wave functions are:

$$\psi_1{'} = 0\psi_1 + 1\psi_2$$
$$\psi_2{'} = -1\psi_1 + 0\psi_2$$

The diagonal elements in the transformation matrix, a_{11} and a_{22}, are both zero, as discussed earlier in Chapter 4 and eq. (4.13). All the characters for this degenerate pair are given in Table 5.2. Comparison of Table 5.2 with the character table for D_{4h} shows that the p_z orbital belongs to species a_{2u} and that the other two p orbitals belong jointly to e_u.

The atomic orbitals on each separate chloride ion do not possess the symmetry of D_{4h}, and accordingly these ligand orbitals must be combined into group orbitals. Consider first the p_z orbitals of the four ligands; those of Cl_1 and Cl_3 lie on the x axis and those of Cl_2 and Cl_4 on the y axis with the positive lobes all pointing toward the metal. The bonds that these orbitals make with metal orbitals will be symmetrical with respect to the metal-ligand axes, and hence will be sigma bonds; consequently these p_z orbitals of the Cl atoms will be referred to as σ_1 to σ_4. The four group orbitals of D_{4h} symmetry, appropriate for combination in sigma bonds with the metal orbitals, can be generated by inspection. These normalized ligand group orbitals (GO) and their symmetry in D_{4h} are:

$$\phi(a_{1g}) = \frac{1}{2}(\sigma_1 + \sigma_2 + \sigma_3 + \sigma_4) \qquad a_{1g}$$

$$\phi(b_{1g}) = \frac{1}{2}(\sigma_1 - \sigma_2 + \sigma_3 - \sigma_4) \qquad b_{1g}$$

$$\left. \begin{array}{l} \phi(e_u) = \dfrac{1}{\sqrt{2}}(\sigma_1 - \sigma_3) \\[2mm] \phi(e_u) = \dfrac{1}{\sqrt{2}}(\sigma_2 - \sigma_4) \end{array} \right\} \qquad e_u$$

These orbitals are represented pictorially in Fig. 5.4. The first is involved in sigma bonding with the metal s and d_{z^2} orbitals, each of which also has a_{1g} symmetry. There are several ways in which the molecular orbital may be developed. The s and d_{z^2} orbitals may first be hybridized and one of the hybrid orbitals combined with the ligand GO in a bonding orbital, the other hydrid orbital has almost no overlap with the ligand GO, and hence remains as a nonbonding MO. Alternately, and preferably, the two metal orbitals and the ligand GO are all combined with appropriate coefficients denoting the contributions of the two metal orbitals and the ligand GO to the total wave functions. The combination of the ligand GO with the two metal AO's leads to three MO's of the form:

$$\psi[a_{1g}] = C_{M1}\phi[d_{z^2}] + C_{M2}\phi[s] + C_L\phi[\tfrac{1}{2}(\sigma_1 + \sigma_2 + \sigma_3 + \sigma_4)]$$

In the *lowest*-energy MO, which is a strongly σ-bonding orbital, the signs of C_L and C_{M2} are equal, and opposite to that of C_{M1}. The magnitudes of these C depend on the various Coulomb, overlap, and resonance integrals involved and are difficult to assess without extensive numerical computation. Correspondingly, the *highest*-energy orbital of a_{1g} symmetry (σ^*) is strongly antibonding, with C_L and C_{M2} of opposite sign, and C_{M1} of the same sign as C_L. Intermediate in energy is the third orbital, which is made up almost completely of the d_{z^2} and s orbitals of Pt, with C_{M1} and C_{M2} of equal sign, and which sticks out strongly above and below the plane; C_L is small, and thus the orbital is essentially nonbonding.

The b_{1g} ligand GO has the same symmetry as the $d_{x^2-y^2}$ platinum

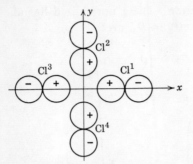

(a) The four $p_z(\sigma)$ orbitals of the chlorine atoms

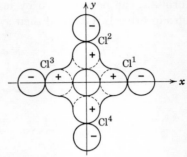

(b) $\phi(a_{1g}) = \frac{1}{2}(\sigma_1 + \sigma_2 + \sigma_3 + \sigma_4)$

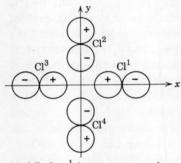

(c) $\phi(b_{1g}) = \frac{1}{2}(\sigma_1 - \sigma_2 + \sigma_3 - \sigma_4)$

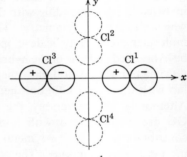

(d) $\phi(e_u) = \frac{1}{\sqrt{2}}(\sigma_1 - \sigma_3)$

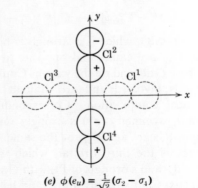

(e) $\phi(e_u) = \frac{1}{\sqrt{2}}(\sigma_2 - \sigma_1)$

Fig. 5.4   (a) The four p_z (σ) orbitals of the chlorine atoms; (b–e) the four GO's generated from (a).

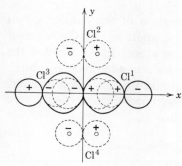

(f) $\psi_I[e_u(\sigma, \pi)] = C_M\phi(p_x) +$

$\quad C_{L1}\phi[\frac{1}{\sqrt{2}}(\sigma_1 - \sigma_3)] -$

$\quad C_{L2}\phi[\frac{1}{\sqrt{2}}(\pi_{2h} - \pi_{4h})]$

(mostly σ)

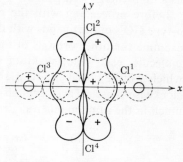

(h) $\psi_I'[e_u(\sigma, \pi)] = C_M\phi(p_x) +$

$\quad C_{L1}\phi[\frac{1}{\sqrt{2}}(\sigma_1 - \sigma_3)] -$

$\quad C_{L2}\phi[\frac{1}{\sqrt{2}}(\pi_{2h} - \pi_{4h})]$

(mostly π)

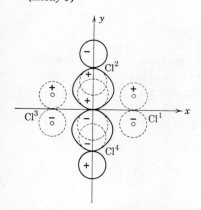

(g) $\psi_{II}[e_u(\sigma, \pi)] = C_M\phi(p_y) +$

$\quad C_{L1}\phi[\frac{1}{\sqrt{2}}(\sigma_2 - \sigma_4)] -$

$\quad C_{L2}\phi[\frac{1}{\sqrt{2}}(\pi_{1h} - \pi_{3h})]$

(mostly σ)

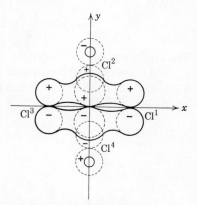

(i) $\psi_{II}'[e_u(\sigma, \pi)] = C_M\phi(p_y) +$

$\quad C_{L1}\phi[\frac{1}{\sqrt{2}}(\sigma_2 - \sigma_4)] -$

$\quad C_{L2}\phi[\frac{1}{\sqrt{2}}(\pi_{1h} - \pi_{3h})]$

(mostly π)

Fig. 5.4 *(f–i)* The e_u MO's.

orbital, and hence these orbitals form a strong σ-bonding MO as well as a high-energy $\sigma^*(b_{1g})$MO.

Before proceeding with the combination of the remaining two σ-type GO's of the ligands with the appropriate metal AO's, we must examine the other orbitals of the ligand atom. In the plane of the molecule, each chlorine atom possesses one additional p orbital, which will be called π_{1h} to π_{4h} and is taken positive in a counterclockwise sense. These orbitals also are combined into GO's, very much in the same manner as the σ orbitals:

$$\phi(a_{2g}) = \frac{1}{2}(\pi_{1h} + \pi_{2h} + \pi_{3h} + \pi_{4h})$$

$$\phi'(b_{2g}) = \frac{1}{2}(\pi_{1h} - \pi_{2h} + \pi_{3h} - \pi_{4h})$$

$$\phi''(e_u) = \frac{1}{\sqrt{2}}(\pi_{1h} - \pi_{3h})$$

$$\phi'''(e_u) = \frac{1}{\sqrt{2}}(\pi_{2h} - \pi_{4h})$$

The remaining MO's formed from the σ-type GO's are of symmetry e_u, and consequently they are linear combinations of all e_u orbitals. Three pairs of degenerate orbitals arise, and each pair has the following form.

$$\psi_{\text{I}}[e_u(\sigma,\pi)] = C_{\text{M}}\phi[\text{p}_x] + C_{\text{L1}}\phi\left[\frac{1}{\sqrt{2}}(\sigma_1 - \sigma_3)\right]$$
$$+ C_{\text{L2}}\phi\left[\frac{1}{\sqrt{2}}(\pi_{2h} - \pi_{4h})\right]$$

$$\psi_{\text{II}}[e_u(\sigma,\pi)] = C_{\text{M}}\phi[\text{p}_y] + C_{\text{L1}}\phi\left[\frac{1}{\sqrt{2}}(\sigma_2 - \sigma_4)\right]$$
$$+ C_{\text{L2}}\phi\left[\frac{1}{\sqrt{2}}(\pi_{1h} - \pi_{3h})\right]$$

One of these pairs will be strongly bonding, one pair strongly antibonding, and the third pair substantially nonbonding. The lowest-energy e_u orbital pair ($\psi_{\text{I}},\psi_{\text{II}}$) is made up mostly of the σ-ligand orbitals; in most MO treatments it is said to be an e_u, σ-bonding orbital and in the crystal field treatment is assigned completely to the ligand. The next highest-energy e_u orbital (ψ_{I}', ψ_{II}') is made up mostly from π contributions, and in most MO treatments is considered essentially an e_u, π-bonding orbital; in the crystal field treatment it is assigned completely to the metal and is the crystal field e_u orbital. Rough sketches of these orbitals are shown in Figs. 5.4f to i. The third e_u doubly degenerate orbital is made up of both σ^* and π^* contributions and is a high-energy antibonding orbital.

Of the two remaining GO's formed from π_{1h} to π_{4h}, $\phi'(b_{2g})$ has the symmetry of the d_{xy} orbital of Pt and hence forms bonding and antibonding MO's with it, and $\phi(a_{2g})$ does not have the symmetry of any metal orbital and hence is nonbonding.

Each ligand atom also has a p orbital normal to the plane of the ion, called π_{1v} to π_{4v}. Their combinations into GO's will be developed below in a more systematic manner to illustrate a general method. Each chlorine atom also has a 2s electron, which has been neglected in the above. They can readily be combined into GO's of the same form and symmetry as the p_z orbitals, and can combine with them to lead to a hybridized chlorine atom.

Although the combination of AO's making up the GO's of appropriate symmetry can be determined by inspection in the relatively simple case of D_{4h}, there is a systematic method for determining the combinations of ligand AO's. For example, suppose we wish to determine the combination of π_v orbitals in Fig. 5.3 which make up four GO's of D_{4h} symmetry. For simplicity, and to use a common shortcut, we utilize the D_4 character table and supply the subscripts g or u by performing the reflection across the σ_h plane.

Step 1. Perform the operations of D_4 on each ligand orbital in turn. It is not necessary to write down the complete transformation matrices, but only the diagonal terms; these are $+1$ if the orbital transforms into itself, -1 if it transforms into minus itself, and 0 if it transforms into another orbital. The sum of these diagonal terms of the transformation matrix for the mutual transformations of the ligands is, of course, the character of the matrix, and hence defines the representation of the transformation. In the bottom half of Chart 5.1 the appropriate diagonal terms are given for the orbitals shown; the last line gives the characters of the reducible representation.

Step 2. Determine the symmetry species or irreducible representations which correspond to the reducible representation. This may be done by inspection or by application of equation (5.6) (see p. 125). In the present case, the reducible representation belongs to species $A_2 + B_2 + E$.

Step 3. Perform the symmetry operations of D_4 on π_{1v} and indicate under the operation the orbital into which π_{1v} transforms (cf. Table 5.3). Multiply the resulting orbitals in turn by the characters of each of the irreducible representations which make up the reducible representation found above (in the

D_4	I	$2C_4$	$C_4{}^2 = C_2{}^z$	$2C_2'$	$2C_2{}^d$
A_1	+1	+1	+1	+1	+1
A_2	+1	+1	+1	−1	−1
B_1	+1	−1	+1	+1	−1
B_2	+1	−1	+1	−1	+1
E	+2	0	−2	0	0
π_{1v}	+1	0	0	−1	0
π_{2v}	+1	0	0	0	0
π_{3v}	+1	0	0	−1	0
π_{4v}	+1	0	0	0	0
Σ	+4	0	0	−2	0

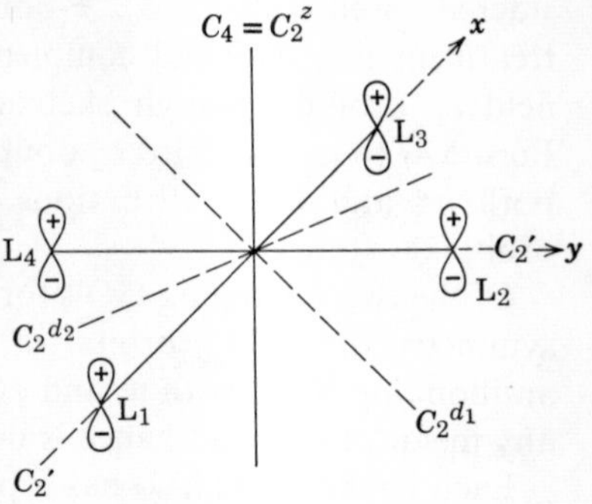

Chart 5.1

present case, A_2, B_2, and E), and sum these products to give the orbital shown in the last column of Table 5.3.

This procedure produces one of the e orbitals, as shown by the product $E \times \pi_{1v}$. In order to get the second e orbital, the same procedure is repeated for $E \times \pi_{2v}$. Thus, the four GO's of appropriate symmetry have been generated; upon normalization and determination of parity (g, u character) they take the form:

$$\phi(a_{2u}) = \frac{1}{2}(\pi_{1v} + \pi_{2v} + \pi_{3v} + \pi_{4v})$$

$$\phi(b_{1g}) = \frac{1}{2}(\pi_{1v} - \pi_{2v} + \pi_{3v} - \pi_{4v})$$

$$\phi(e_u) = \frac{1}{\sqrt{2}}(\pi_{1v} - \pi_{3v})$$

$$\phi'(e_u) = \frac{1}{\sqrt{2}}(\pi_{2v} - \pi_{4v})$$

These orbitals were derived on the basis of the arbitrary assignment of the x and y axes to the Cl^1PtCl^3 and Cl^2PtCl^4 axes. Any other pair of orthogonal lines in the same plane can serve equally well as x and y axes. The bisectors of the previous two axes are particularly convenient, and lead to the same a_{2u} and b_{1g} orbitals, but to

$$\phi''(e_u) = \frac{1}{2}(\pi_{1v} + \pi_{2v} - \pi_{3v} - \pi_{4v}) = \frac{1}{\sqrt{2}}(\phi(e_u) + \phi'(e_u))$$

$$\phi'''(e_u) = \frac{1}{2}(\pi_{1v} - \pi_{2v} - \pi_{3v} + \pi_{4v}) = \frac{1}{\sqrt{2}}(\phi(e_u) - \phi'(e_u))$$

The above is not intended to be a complete MO treatment of $[PtCl_4]^{2-}$; for such a treatment see the last reference at the end of this chapter.

Table 5.3

	I	C_4, C_4'	$C_4{}^2 = C_2{}^z$	$C_2{}^x, C_2{}^y$	$C_2{}^{d1}, C_2{}^{d2}$	Group Orbital
π_{1v}	π_{1v}	π_{2v}, π_{4v}	π_{3v}	$-\pi_{1v}, -\pi_{3v}$	$-\pi_{2v}, -\pi_{4v}$	
A_2	$+1$	$+1, +1$	$+1$	$-1, -1$	$-1, -1$	
$A_2 \times \pi_{1v}$	π_{1v}	π_{2v}, π_{4v}	π_{3v}	$-\pi_{1v}, -\pi_{3v}$	$-\pi_{2v}, -\pi_{4v}$	$2\pi_{1v} + 2\pi_{2v} + 2\pi_{3v} + 2\pi_{4v}$
B_2	$+1$	$-1, -1$	$+1$	$-1, -1$	$+1, +1$	
$B_2 \times \pi_{1v}$	π_{1v}	$-\pi_{2v}, -\pi_{4v}$	π_{3v}	π_{1v}, π_{3v}	$-\pi_{2v}, -\pi_{4v}$	$2\pi_{1v} - 2\pi_{2v} + 2\pi_{3v} - 2\pi_{4v}$
E	$+2$	$0, 0$	-2	$0, 0$	$0, 0$	
$E \times \pi_{1v}$	$2\pi_{1v}$	—	$-2\pi_{3v}$	—	—	$2(\pi_{1v} - \pi_{3v})$
π_{2v}	π_{2v}	π_{3v}, π_{1v}	π_{4v}	$-\pi_{4v}, -\pi_{2v}$	π_{1v}, π_{3v}	
$E \times \pi_{2v}$	$2\pi_{2v}$	—	$-2\pi_{4v}$	—	—	$2(\pi_{2v} - \pi_{4v})$

Finally, consider SF_6 and CCl_4; the S and C atoms lie on all symmetry elements, and their orbitals transform like the various species of their point groups, O_h and T_d. The F and Cl orbitals are assembled into group orbitals as shown in Tables 5.4 and 5.5. It may be noted that, for convenience, a different coordinate system is applied to each ligand atom, such that the positive z direction of each ligand points toward the central atom. The 21st and 27th-order equations obtained without factoring reduce to equations of much lower order, no greater than the third or fourth, by this procedure for CCl_4 and SF_6, respectively. A phenomenal saving of labor!

5.2 NUMBER OF NORMAL VIBRATIONS

In Chapter 4 we saw that any given molecule has $3n - 6$ (or, if linear, $3n - 5$) normal vibrations, each potentially giving rise to a fundamental frequency in the infrared, and that these vibrations could be classified according to symmetry species. In this section we shall examine the way in which the symmetry properties alone permit us to predict how many of the normal vibrations belong to any one given symmetry species, and hence how many fundamental frequencies may be anticipated. This information may be of tremendous importance: Vibrations belonging to *some* symmetry species

Table 5.4   The Group Orbitals of CCl$_4$, Point Group T_d

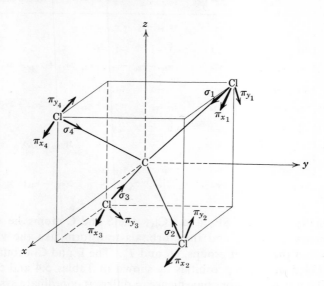

T_d Carbon Orbitals		Group Orbitals (all $\times \frac{1}{4}$)
A_1	s	$2(\sigma_1 + \sigma_2 + \sigma_3 + \sigma_4)$
A_2		
E		$[\pi_{x_1} + \pi_{x_2} + \pi_{x_3} + \pi_{x_4} - \sqrt{3}(\pi_{y_1} + \pi_{y_2} + \pi_{y_3} + \pi_{y_4})]$
		$[\pi_{y_1} + \pi_{y_2} + \pi_{y_3} + \pi_{y_4} + \sqrt{3}(\pi_{x_1} + \pi_{x_2} + \pi_{x_3} + \pi_{x_4})]$
T_1		$[\pi_{y_2} + \pi_{y_4} - \pi_{y_3} - \pi_{y_1} + \sqrt{3}(\pi_{x_1} + \pi_{x_3} - \pi_{x_2} - \pi_{x_4})]$
		$(\pi_{y_1} + \pi_{y_2} - \pi_{y_3} - \pi_{y_4})$
		$[\pi_{y_2} + \pi_{y_3} - \pi_{y_1} - \pi_{y_4} + \sqrt{3}(\pi_{x_2} + \pi_{x_3} - \pi_{x_1} - \pi_{x_4})]$
T_2		$2(\sigma_1 + \sigma_3 - \sigma_2 - \sigma_4)$
	p_x	$[\pi_{x_4} + \pi_{x_2} - \pi_{x_1} - \pi_{x_3} + \sqrt{3}(\pi_{y_4} + \pi_{y_2} - \pi_{y_1} - \pi_{y_3})]$
	p_y	$2(\sigma_1 + \sigma_2 - \sigma_3 - \sigma_4)$
	p_z	$(\pi_{x_1} + \pi_{x_2} - \pi_{x_3} - \pi_{x_4})$
		$2(\sigma_1 + \sigma_4 - \sigma_2 - \sigma_3)$
		$[\pi_{x_3} + \pi_{x_2} - \pi_{x_1} - \pi_{x_4} + \sqrt{3}(\pi_{y_4} + \pi_{y_1} - \pi_{y_2} - \pi_{y_3})]$

The coordinate system on C is right-handed, those on Cl's are left-handed.
Cf. M. Wolfsberg and L. Helmholz, *J. Chem. Phys.*, **20,** 837 (1952).

Table 5.5   The Group Orbitals of SF_6, Point Group O_h

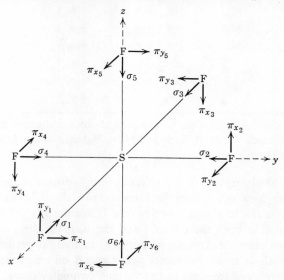

O_h	Sulfur Orbitals	Ligand Group Orbitals
A_{1g}	3s	$(1/\sqrt{6})(\sigma_1 + \sigma_2 + \sigma_3 + \sigma_4 + \sigma_5 + \sigma_6)$
A_{1u}		
A_{2g}		
A_{2u}		
E_g	$3d_{z^2}$	$(1/2\sqrt{3})(2\sigma_5 + 2\sigma_6 - \sigma_1 - \sigma_2 - \sigma_3 - \sigma_4)$
	$3d_{x^2-y^2}$	$\frac{1}{2}(\sigma_1 - \sigma_2 + \sigma_3 - \sigma_4)$
E_u		
T_{1g}		$\frac{1}{2}(\pi_{y_1} - \pi_{x_5} + \pi_{x_3} - \pi_{y_6})$
		$\frac{1}{2}(\pi_{x_2} - \pi_{y_5} + \pi_{y_4} - \pi_{x_6})$
		$\frac{1}{2}(\pi_{x_1} - \pi_{y_2} + \pi_{y_3} - \pi_{x_4})$
T_{1u}	$3p_x$	$(1/\sqrt{2})(\sigma_1 - \sigma_3)$
		$\frac{1}{2}(\pi_{y_2} + \pi_{x_5} - \pi_{x_4} - \pi_{y_6})$
	$3p_y$	$(1/\sqrt{2})(\sigma_2 - \sigma_4)$
		$\frac{1}{2}(\pi_{x_1} + \pi_{y_5} - \pi_{y_3} - \pi_{x_6})$
	$3p_z$	$(1/\sqrt{2})(\sigma_5 - \sigma_6)$
		$\frac{1}{2}(\pi_{y_1} + \pi_{x_2} - \pi_{x_3} - \pi_{y_4})$

Table 5.5   (*Continued*)

T_{2g}	$3d_{xz}$	$\frac{1}{2}(\pi_{y_1} + \pi_{x_5} + \pi_{x_3} + \pi_{y_6})$
	$3d_{yz}$	$\frac{1}{2}(\pi_{x_2} + \pi_{y_5} + \pi_{y_4} + \pi_{x_6})$
	$3d_{xy}$	$\frac{1}{2}(\pi_{x_1} + \pi_{y_2} + \pi_{y_3} + \pi_{x_4})$
T_{2u}		$\frac{1}{2}(\pi_{y_2} - \pi_{x_5} - \pi_{x_4} + \pi_{y_6})$
		$\frac{1}{2}(\pi_{x_1} - \pi_{y_5} - \pi_{y_3} + \pi_{x_6})$
		$\frac{1}{2}(\pi_{y_1} - \pi_{x_2} - \pi_{x_3} + \pi_{y_4})$

The coordinate system of the S atom is right-handed, those of the F atoms are left-handed. Cf. H. B. Gray and N. A. Beach, *J. Am. Chem. Soc.*, **85**, 2922 (1963).

give rise to absorption in the infrared (are "*infrared active*"); those belonging to *some* species (not necessarily all different from the above) give rise to Raman spectra (are "*Raman active*"); and *others* are inactive in both the infrared and the Raman. In a subsequent section we examine how this activity or inactivity is predicted.

The analysis to be performed will depend on assignments of *all* the $3n$ degrees of freedom to the various symmetry types. We have already developed the symmetry types of the three translations and of the three (or, in the linear case, two) rotations. These six (or five) degrees of freedom are called *nongenuine vibrations*. A genuine molecular vibration involves a change in the *relative* position of atoms only; simple translations and rotations preserve the relative positions of the atoms with respect to each other, and hence are nongenuine vibrations. The number of nongenuine vibrations must be subtracted from the total number of vibrations in each symmetry species to give the number of genuine vibrations of each symmetry type.

Take the example of water again. We have seen that the three atoms permit $3 \times 3 - 6 = 3$ normal modes, which have been shown in Fig. 4.11. How can we determine that just two of the fundamentals must belong to A_1 and one to B_2, as pointed out above? First we recognize that the two hydrogen atoms are equivalent; they form one *set* of atoms. The oxygen atom, which lies on all symmetry elements, belongs to a set by itself, and has no equivalent. In general, in the point group C_{2v}, an atom or a point lying on no symmetry element must belong to a set of four equivalent atoms, as it is most easily verified by examination of a stereographic projection (cf. Fig. 5.5). Any atom, or point, lying in either of the planes but not on the axis must occur as part of a set of two. In this point group, as in all others, any atom which lies on *all* symmetry elements is unique and forms a set by itself.

In the simple (nondegenerate) point groups, if the displacement of one atom of a set in a given vibration (genuine or nongenuine) is specified, the displacements of the other atoms are given by the symmetry. Let us return to the water molecule. For a vibration of species A_1 or B_2, the motion of the H atoms must lie in the molecular plane, since the motion must be symmetric with respect to reflection at this plane; motion in a plane can always be resolved into two orthogonal components. Hence there are two degrees of freedom of motion in a plane, and this set of atoms contributes two degrees of freedom. For vibrations of species A_2 or B_1, the motion is out of the molecular

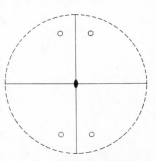

Fig. 5.5 Stereographic projection of general point in point group C_{2v}.

plane, and the direction of motion of the H's is completely specified, normal to the plane; hence the two atoms jointly contribute one degree of freedom to each of these species. In A_1, the oxygen atom must move along the symmetry axis and thus contributes one degree of freedom; in B_1 and B_2, it must move normal to the axis and in the plane of symmetry xz for B_1, yz for B_2, and hence contributes one degree of freedom to each. Thus, A_1 has three degrees of freedom (two from H's, one from O), A_2 has one (from the H's) B_2 three (two from H's, one from O), and B_1 two (one each from H's and O) for a total of nine degrees of freedom, as required. These degrees of freedom include the three translations (A_1, B_1, and B_2) and the three rotations (A_2, B_1, and B_2), and this leaves, for vibrations, just two in A_1, one in B_2, and none in A_2 or B_1.

There is a more formal recipe for determining the symmetry species of the $3n$ vibrations. Each atom is assigned a set of three orthogonal displacement vectors, one parallel to each Cartesian coordinate axis, so that the vectors on each atom are parallel. Then in turn each symmetry operation appropriate to the molecule is performed on these vectors. In performing these operations, we assume that the atoms remain fixed and only the vectors are shifted. The *character* of the transformation matrix for each operation is determined, and hence a reducible representation is developed. Finally, the direct sum of the irreducible representations corresponding to the reducible representation is determined. The method will be illustrated by the use of the water molecule.

The coordinate system is set up as shown in Fig. 5.6; the coordinates resulting from a transformation will be indicated by primes. The transformation arising from the C_2^z operation is represented by the transformation matrix shown in

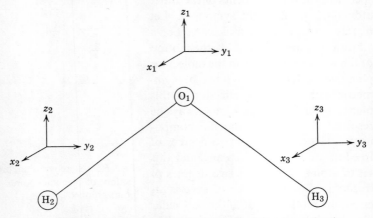

Fig. 5.6   The set of displacement vectors for the water molecule.

Table 5.6a. Note that only the coordinate vectors on the oxygen atom contribute to the character since this atom is the only one transformed into itself by the operation. Upon performing the σ_{xz} operation, again only the coordinates on the oxygen need be considered, and the relevant partial transformation matrix shown in Table 5.6b gives a character of $+1$. Now on σ_{yz}, the coordinates of all three atoms remain on the same atom, and Table 5.6c gives the appropriate matrix, which has a character of $+3$. Incidentally, instead of a 9×9 matrix, a 3×3 submatrix could have been used, since all three atoms are affected in an identical way by the operation σ_{yz}. The character of the submatrix is $+1$, and multiplication by 3 gives the value of $+3$ found for the 9×9 matrix. Finally, under the operation I, the coordinates of each atom remain unchanged; the character of the 3×3 submatrix is $+3$ and of the 9×9 matrix $+9$. Accordingly, the characters of the reducible representation are:

C_{2v}	I	C_2^z	σ_{xz}	σ_{yz}
	9	-1	$+1$	$+3$

Now we must determine the sum of the irreducible representations that correspond to the above reducible representation. Application of equation (5.6) on p. 125 gives:

$$3A_1 + A_2 + 2B_1 + 3B_2$$

From these representations we may subtract the symmetry species of the three translations (see Table A1.2) (A_1, B_1, and B_2) and of the three rotations (A_2, B_1, and B_2) leaving $2A_1 + B_2$ identical with the three vibrations deduced above by considering the degrees of freedom.

Table 5.6 Matrices Expressing the Effect of the (a) $C_2{}^z$, (b) σ_{xz}, and (c) σ_{yz} Operations on the Cartesian Coordinates for H_2O Shown in Fig. 5.6.

$C_2{}^z$

	x_1	y_1	z_1	x_2	y_2	z_2	x_3	y_3	z_3
x_1'	−1	0	0	0	0	0	0	0	0
y_1'	0	−1	0	0	0	0	0	0	0
z_1'	0	0	1	0	0	0	0	0	0
x_2'	0	0	0	0	0	0	−1	0	0
y_2'	0	0	0	0	0	0	0	−1	0
z_2'	0	0	0	0	0	0	0	0	1
x_3'	0	0	0	−1	0	0	0	0	0
y_3'	0	0	0	0	−1	0	0	0	0
z_3'	0	0	0	0	0	1	0	0	0

(a)

σ_{xz}

	x_1	y_1	z_1
x_1'	1	0	0
y_1'	0	−1	0
z_1'	0	0	1

(b)

σ_{yz}

	x_1	y_1	z_1	x_2	y_2	z_2	x_3	y_3	z_3
x_1'	−1	0	0	0	0	0	0	0	0
y_1'	0	1	0	0	0	0	0	0	0
z_1'	0	0	1	0	0	0	0	0	0
x_2'	0	0	0	−1	0	0	0	0	0
y_2'	0	0	0	0	1	0	0	0	0
z_2'	0	0	0	0	0	1	0	0	0
x_3'	0	0	0	0	0	0	−1	0	0
y_3'	0	0	0	0	0	0	0	1	0
z_3'	0	0	0	0	0	0	0	0	1

(c)

Next let us examine the hexacoordinated complex shown in Fig. 5.7, which also belongs to C_{2v}. The ammino groups are so oriented that the upper H atom lies in the NBrFM plane. M, Br, and F are each one set, each lying on all elements. The two Cl's, the two N's, and the two upper H's are three sets lying on one plane or the other.

The four lower H's are one further set lying on no element of symmetry. In each species the lower H's, representing general points, have three degrees of freedom. Any motion of the N's and upper H's in the x direction is antisymmetric with respect to reflection from the yz plane in which these atoms lie. Hence they have only *one* degree of freedom in each of the symmetry species A_2 and B_1, which have this same symmetry behavior. The other *two* degrees of freedom of these two sets of atoms, namely motion in the y and z direction (or

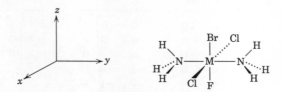

Fig. 5.7   A hexacoordinated complex of point group C_{2v}.

any combination of the two), lie in the yz plane; hence they are symmetric with respect to reflection from it, and contribute *two* degrees of freedom to each of the species A_1 and B_2, which also are symmetric with respect to the same operation. Similarly, the set of Cl atoms lying in the xz plane has one degree of freedom in A_2 and B_2 (motion in the y direction), two in A_1 and B_1 (motion in the x and z directions). The M, Br, and F have one degree of freedom each in A_1 (along C_2), B_1 (perpendicular to C_2 in $\sigma_v{}^{xz}$), and B_2 (perpendicular to C_2 and in $\sigma_v{}^{yz}$) but none in A_2. Hence the total number of degrees of freedom is given by:

$$
\begin{aligned}
A_1: &\quad 3 \times 1 + 2 \times 2 + 2 \times 1 + 1 \times 3 = 12 \\
A_2: &\quad 3 \times 1 + 1 \times 2 + 1 \times 1 \qquad\qquad\; = 6 \\
B_1: &\quad 3 \times 1 + 1 \times 2 + 2 \times 1 + 1 \times 3 = 10 \\
B_2: &\quad 3 \times 1 + 2 \times 2 + 1 \times 1 + 1 \times 3 = 11 \\
\text{Total:} &\quad 12 \times 1 + 6 \times 2 + 6 \times 1 + 3 \times 3 = \overline{39}
\end{aligned}
$$

These equations have been derived for the specific complex shown in Fig. 5.7, but they can readily be generalized to apply to any molecule of the same point group, C_{2v}. If we designate the number of sets of atoms lying on no element of symmetry by m, the number of sets on all elements by m_0, and the number on one or the other of the two planes by $m_{\sigma(xz)}$ and $m_{\sigma(yz)}$, this becomes:

$$A_1: \quad 3m + 2m_{\sigma(xz)} + 2m'_{\sigma(yz)} + m_0$$

$$A_2: \quad 3m + \quad m_{\sigma(xz)} + \quad m'_{\sigma(yz)}$$

$$B_1: \quad 3m + \quad m_{\sigma(xz)} + 2m'_{\sigma(yz)} + m_0$$

$$B_2: \quad 3m + 2m_{\sigma(xz)} + \quad m'_{\sigma(yz)} + m_0$$

Now we have seen in the past that the three translations and the three rotations are six degrees of freedom which are not genuine vibrations; since these six degrees belong, one to A_1, one to A_2, two to B_1 and B_2 each, they must be subtracted from the above totals, giving for the number of normal vibrations for any molecule in the point group C_{2v}:

$$A_1: \quad 3m + 2m_{\sigma(xz)} + 2m'_{\sigma(yz)} + \quad m_0 - 1$$

$$A_2: \quad 3m + \quad m_{\sigma(xz)} + \quad m'_{\sigma(yz)} \qquad - 1$$

$$B_2: \quad 3m + \quad m_{\sigma(xz)} + 2m'_{\sigma(yz)} + \quad m_0 - 2$$

$$B_1: \quad 3m + 2m_{\sigma(xz)} + \quad m'_{\sigma(yz)} + \quad m_0 - 2$$

$$\text{Total:} \quad 12m + 6m_{\sigma(xz)} + 6m'_{\sigma(yz)} + 3m_0 - 6 \qquad (5.2)$$

It is interesting to check the totals. There are four m atoms of sets of type m, each with three degrees of freedom, and therefore $12m$ degrees of freedom; similarly $6m_{\sigma(xz)}$, $6m_{\sigma}'{}_{(yz)}$, and $3m_0$, which are just the sums of the appropriate columns in eq. (5.2). In other words, what we have done is to distribute the total number of degrees of freedom of the molecule between the different symmetry species. Analogous general equations can readily be obtained for other nondegenerate point groups and are summarized in Appendix 2.

In the case of degenerate point groups, the problems are unchanged as regards the nondegenerate species; however, in degenerate species a displacement of one atom of a set no longer unequivocally defines the displacements of the other atoms of the set. Consider as an example ammonia, in point group C_{3v}. The nitrogen atom contributes one degree of freedom to A_1 (motion along the axis), none to A_2 (since motion out of the axis is degenerate), and two to E; since these two are degenerate, this implies a contribution of *one* normal mode. Motion of the N atoms in the σ_v planes contributes two degrees of freedom to A_1 (motion out of these planes), one to A_2. But to any one displacement of one of the H atoms (which may be in any of three directions), there correspond, in species E, two possible displacements of the other two H atoms. Hence the H atoms contribute six degrees of freedom or three normal modes to E. A set of atoms not on any of the symmetry elements, that is, in a general position (such a set consists of six equivalent atoms, see the stereographic projection in Fig. 3.11c) would contribute twelve degrees of freedom or six normal modes to E. The analogous arguments can be made for all point groups, although with increasing difficulty, and lead to the general equations given in Appendix 2.

5.3 SELECTION RULES AND POLARIZATION. THE DIRECT PRODUCT

One of the most interesting and important applications of symmetry considerations in chemistry is the determination of selection rules. Application of these rules specifies whether a given transition between two states is allowed or forbidden. This type of application is possible in all types of spectra, but of particular importance in electronic (ultraviolet and visible) and vibrational (infrared and Raman) spectra, although also in rotational (microwave) and nuclear magnetic resonance spectra. The determination of selection rules depends on the evaluation of the intensity of a given transition.

The treatments of the selection rules for electronic and vibrational spectra are so similar that the two can readily be combined. A selection rule is simply an expression of the statement that the intensity of a given transition, as calculated by an approximate theory, is zero; in this case the transition is called *forbidden*, and, according to the *approximate* theory, should not be observed. Actually, any theory we use makes approximations, and many forbidden transitions are observed, but invariably with relatively low intensity. Some selection rules do not depend on symmetry considerations. Examples are rules forbidding the transitions between states of different multiplicity, that is, states of molecules with different total electronic spin, and between states that require jumps of several electrons. Such selection rules will not be further discussed.

The heart of the selection rules is the *transition moment*, an integral the square of which is called the dipole strength D and determines, except for fundamental constants, the intensity of any transition.[1] The transition moment $\sqrt{D}$ for a transition between two states of a molecule described by the wave functions Ξ_i (initial) and Ξ_f (final) is given by

$$\sqrt{D} = \int \Xi_i \mathbf{M} \Xi_f \, d\tau \tag{5.3}$$

$\mathbf{M}$ is the dipole moment vector with components $M_x = \Sigma \, e_j x_j$, $M_y = \Sigma \, e_j y_j$, $M_z = \Sigma \, e_j z_j$, where e_j is the charge of particle j having coordinates x_j, y_j, and z_j; the summation extends over all nuclei and electrons. The wave functions Ξ are functions of the coordinates of all these particles, and the integration extends over all space in all these particles. In the first—and fairly good—approximation Ξ may be factored into a product of a series of functions, $\Psi \cdot X \cdot R$, where Ψ is the electronic wave function, a function of the coordinates of all the electrons; X is the vibrational wave function, a function of the coordinates of the nuclei; and R is a rotational wave function, a function of the angles fixing the rotational axes in space.

To the approximation used here, eq. (5.3) then reduces, for electronic spectra, to:

$$\sqrt{D} = \int \Psi_i \mathbf{M} \Psi_f \, d\tau \tag{5.4a}$$

and for vibrational (infrared) spectra to:

$$\sqrt{D} = \int X_i \mathbf{M} X_f \, d\tau \tag{5.4b}$$

X is the product of a series of one-dimensional vibrational functions, χ, one for each fundamental; similarly, but only approximately, Ψ is the product of a series of one-electron functions, ψ, and eqs (5.4) reduce to

$$\sqrt{D} = e \int \psi_i \mathbf{M} \psi_f \, d\tau \tag{5.5a}$$

$$\sqrt{D} = e \int \chi_i \mathbf{M} \chi_f \, d\tau \tag{5.5b}$$

---

[1] Strictly, this statement applies only to electric dipole transitions. Transitions of higher order (quadrupole, etc.) and magnetic dipole and higher-order transitions may also occur, but with much lower intensity, and are here completely neglected.

Each of these integrals can be resolved into three components, according to the three components of $\mathbf{M} = M_x + M_y + M_z$, which have the symmetry properties of x, y, and z, or of the translations (velocity vectors) in the x, y, and z directions.

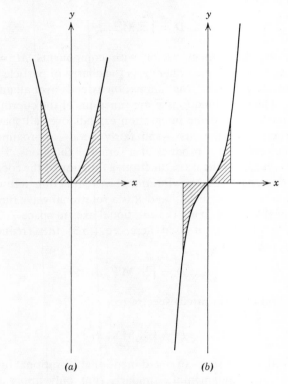

(a)   *(b)*

Fig. 5.8   Even *(a)* and odd *(b)* functions $y(x)$.

Next, let us examine how the symmetry properties help determine whether integrals of the type of eqs. (5.5) vanish. It is a well-known fact of mathematics that $\int y\,dx$ represents the area under the curve if y is plotted against x, areas below the abscissa being counted negative. Now, assume functions $y(x)$ such that y^2 is symmetric in x; that is, $y^2(x) = y^2(-x)$. One such function $y = x^2$ is shown in Fig. 5.8a. In this case $y(x) = y(-x)$ (that is, for two identical values of x which differ only in sign, y has the same value); such a function is called *even*. The (shaded) area under the parabola can be expressed by the integral

$$\int_{-a}^{+a} y \, dx = 2 \int_{0}^{a} y \, dx$$

since the areas under the two halves of the curve both are above the abscissa and are equal. A plot of the equation $y = x^3$ is shown in Fig. 5.8b, where it can be seen that for every positive value of x there is a positive value of y, and for every negative value of x a corresponding negative value of y. Here $y(x) = -y(-x)$; such a function is called an *odd* function. Integration of $y = x^3$ gives

$$\int_{-a}^{+a} y \, dx = 0$$

since the (shaded) areas under the curve for positive and negative x are equal but above and below the abscissa, respectively, and thus cancel. The product of an even function and an odd function is always odd, while the product of two even or of two odd functions is always even. The integral over all space of an odd function always vanishes identically, whereas the integral over all space of an even function may or may not vanish, but generally does not.

Then, any of the component integrals of eqs. (5.5) vanish whenever the integrand is antisymmetric with respect to *any* symmetry element. In other words, a necessary condition for a transition to be *allowed*, that is, not to be forbidden, is that the integrand be totally symmetric (belong to the species A, A_1, A_g, or A' of the appropriate point group; that is, have a character of $+1$ for each symmetry operation). Since we know the symmetry of each of the ψ, χ, and the components of **M**, it then remains to determine the symmetry characteristics of a product of functions belonging to a given symmetry species.

The first conclusion readily arrived at relates to molecules with a center of symmetry. The product of any two functions which are both symmetric (g) or both antisymmetric (u) with respect to a center is g; the product of one g and one u function is itself u. Now, all components of **M** are necessarily u (if a center exists), and thus all three components of the integral eqs. (5.5) vanish unless the wave function product $\psi_i\psi_f$ or $\chi_i\chi_f$ is u. This is often stated as the selection rule $g \leftrightarrow u$ but $g \not\leftrightarrow g$, $u \not\leftrightarrow u$, that is, only $g \leftrightarrow u$ transitions are allowed. This is a rather strong selection rule and holds rigidly in the infrared; in the UV and visible, transitions violating this rule are quite weak.

Nondegenerate Point Groups. For the nondegenerate symmetry species, obtaining the symmetry characteristics of the product of two functions is a rather simple process. It is intuitively obvious that the product of two antisymmetric functions is a symmetric one, just as $(-1) \times (-1)$ is $+1$. Since the character of an antisymmetric function is -1 and that of a symmetric one $+1$, all that is needed is the multiplication of the characters. Thus for water, discussed above, we had orbitals of species a_1, b_1, and b_2 (cf. Table 5.1), all of which were occupied. Of these, one a_1 and b_1 represent lone pairs. Vacant antibonding orbitals $a_1{}^*$ and $b_2{}^*$ correspond to the other a_1 and b_2 orbital. Let us consider the excitation of one of the lone-pair electrons to one of the antibonding orbitals. We have then four possible transitions, which may be described, in approximate order of decreasing wavelength,

$$b_1 \rightarrow b_2{}^*; \quad b_1 \rightarrow a_1{}^*; \quad a_1 \rightarrow b_2{}^*; \quad a_1 \rightarrow a_1{}^*$$

Let us then inquire into the symmetry of the integrand of eq. (5.5a). Multiplying together the characters of b_2 and b_1, we obtain for the product $\psi_i \psi_f$ the characters $+1 \times +1 = +1$, $-1 \times -1 = +1$, $-1 \times +1 = -1$, and $+1 \times -1 = -1$, that is, the characters of A_2. Multiplying this product in turn by the characters of $x(b_1)$, $y(b_2)$, and $z(a_1)$, we find that none of the three partial integrands transforms as a_1, none of them is totally symmetric, and the transition is forbidden.

If we do the multiplication for $b_1 \times a_1{}^*$, $a_1 \times b_2{}^*$, and $a_1 \times a_1{}^*$, we find the three products, respectively, to transform as B_1, B_2, and A_1. If we multiply each of these successively by x, y, and z, we find of the nine partial integrals that $b_1 \times a_1{}^* \times x$, $a_1 \times b_2{}^* \times y$, and $a_1 \times a_1{}^* \times z$, and only these, are totally symmetric. Since, then, there is one nonvanishing term in the intensity of each of the last three transitions, each of them is allowed.

The information we have just obtained is even more valuable than just the determination of allowed or forbidden nature of the transition. The theory so far developed applies to ordinary electromagnetic radiation. If the incident radiation is polarized (say by passage through a Nicol prism), the dipole moment vector **M** in eq. (5.3) (and subsequent equations) must be replaced by only its component in the direction of polarization of the light. Since, normally, molecules are oriented randomly with respect to the incident light, all possible relations between the axis of the light and of the molecules occur, and the intensity is unaffected. But by use of a single crystal or of a monomolecular film of oriented molecules, it is possible to achieve a fixed relation between the axis (or plane) of the molecule and the direction of polarization of the incident light.

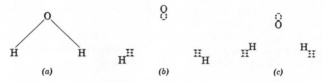

Fig. 5.9   The totally symmetric stretching vibration of H_2O.

Consequently it is of interest to consider the absorption of light polarized in different directions with respect to the molecular symmetry elements.

The argument shown above indicates then that the $b_1 \rightarrow a_1{}^*$ transition absorbs only if the x axis of the molecule coincides with the direction of polarization of the exciting light; or in other words, this transition is x *polarized*. Similarly, the $a_1 \rightarrow b_2{}^*$ transition is y and the $a_1 \rightarrow a_1{}^*$ transition z *polarized*.

The discussion of infrared transitions is exactly analogous, except that we must first determine the symmetry of the wave functions. We have determined classically the symmetry species to which a vibration belongs. Vibrations, however, do not behave classically; they must be described by wave functions, which are functions of the coordinates of the atoms, or better, of the symmetry coordinate of the molecule and the vibrational quantum number.

The symmetry coordinate, usually called ξ, is a single coordinate which measures the progress of the molecule along the vibration path. As pointed out above, the normal vibrations are so chosen that all atoms move in phase; in other words, all atoms simultaneously pass through the equilibrium configuration (cf. Fig. 5.9a), reach the extreme end of the motion at one time when they turn around (Fig. 5.9b), again simultaneously pass through the equilibrium configuration to reach the other extreme all at one time (Fig. 5.9c). The coordinate, which measures the progress along this path, of necessity transforms under symmetry operations like the vibration, that is, like the symmetry species of the vibration.

Now the vibrational wave function of any vibration v_i and quantum number v_j is given by:

$$\chi_{ij} = N_{ij} H_j(\xi_i) e^{-\alpha_i^2 \xi_i^2 / 2}$$

$H_j(\xi_i)$ is a polynomial of order j; for $j = 0$ ($v_j = 0$) it is just 1; for $j = 1$ ($v_j = 1$) it is just ξ_i. The exponential is seen to depend on ξ_i^2, and hence is totally symmetric. Then χ_{i0}, the vibrational wave function for the ground state ($v_0 = 0$) is just $N_{i0} e^{-\alpha_i^2 \xi_i^2 / 2}$ and consequently totally symmetric; for the first excited state $v_1 = 1$, $\chi_{i1} = N_{i1} \xi_i e^{-\alpha_i^2 \xi_i^2 / 2}$ and hence transforms as ξ_i, that is, as the vibration. The $H_j(\xi_i)$ contain only even powers of ξ_i if j is even, only odd powers if j is odd. Consequently, the χ_{ij} in the nondegenerate species are totally symmetric for even j, and transform as the vibration for odd j.

In water we had three vibrations, two of species A_1 and one of B_2. In the A_1 fundamentals all wave functions transform as A_1: in the B_2 vibration, even-numbered wave functions transform as A_1, odd-numbered ones as B_2. In vibrational, as in electronic, spectra, selection rules not directly based on molecular symmetry are operative. Two of these state that, to a first approximation, $\Delta v = 1$, that is, the vibrational quantum number changes only by one, and only one vibration changes quantum number in any one transition. Hence, in the intensity-determining integral

$$\sqrt{D} = \int \chi_i \mathbf{M} \chi_f \, d\tau$$

$\chi_i \chi_f$ transforms as $A_1 \times A_1 = A_1$ for the A_1 vibrations, as $A_1 \times B_2 = B_2$ for the B_2 vibration. Since $\mathbf{M}$ has components of both $A_1(z)$ and $B_2(y)$ symmetry, both vibrations are infrared active, but following what was said above, two (of species A_1) are z polarized, one (of B_2) y polarized. In the point group C_{2v}, only vibrations of species A_2 are forbidden; while H_2O has no such vibrations, they occur frequently in more complicated molecules. The complex of Fig. 5.7 has five of them, as we calculated above.

The selection rule which makes these transitions forbidden may be stated in three ways: (*a*) A transition is forbidden if none of the three components of the intensity integral is nonvanishing; or (*b*) a transition is forbidden if none of the components of $\mathbf{M}$ has the same symmetry as the product of initial and final wave function. Or, again, for the infrared only, (*c*) a vibration is inactive if the dipole moment of the molecule does not change during the vibration. This last statement can be shown to be equivalent to the other two, but the proof will not be given here.

The selection rules that $\Delta v = 1$, and that only one vibration changes in a given transition, are approximations which are valid only if the vibration is *harmonic*, that is, if the potential energy curve is a parabola, or if the restoring force is proportional to the displacement from the equilibrium position. Actual vibrations, however, are only nearly harmonic, and hence both selection rules are violated, although the resulting transitions are of low intensity. When $\Delta v > 1$, we speak of *overtones*, and when several vibrational quantum numbers change at once, of *combination bands*. Both overtones and combination bands, however, are subject to the same symmetry selection rules as the usual transitions. Consequently, it is possible to predict that a certain overtone *cannot* occur if the product $\chi_i \chi_f$ (with $\Delta v > 1$) does not belong to the same symmetry species as at least one of the components of **M**. Similarly, if the product of the initial and final vibrational wave functions $X_i X_f$, each of which is the product of two (or more) wave functions for one vibration, for a combination band does not transform like any component of **M**, the combination cannot be observed. Unfortunately, the reverse is not true, and overtones and combination bands which are *not* forbidden by this criterion may still have sufficiently low intensity not to be observed.

Another similar combination may occur between electronic and vibrational functions. For instance, we saw above that the $b_1 \to b_2*$ transition of water is forbidden. If this, however, is accompanied by a vibrational quantum jump of the B_2 vibration, we must resort back to eq. (5.3). The Ξ, a so-called vibronic function, may now be taken as the product of electronic and vibrational wave functions. $\Xi_i = b_1 \times (a_1)$ is of species B_1; $\Xi_j = b_2 \times (b_2)$ of species A_1, if the vibrational jump is from an even to an odd level; $\Xi_i = b_1 \times (b_2)$ of A_2, $\Xi_f = b_2 \times (a_1)$ of B_2 for an odd-to-even jump. In either case $\Xi_i \Xi_j$ is B_1, and the vibronic transition is allowed and x polarized.

Degenerate Point Groups. We may now proceed to a consideration of the same problems for degenerate species. Nothing new arises as far as products of two quantities (wave functions or components of **M**) belonging to nondegenerate, or one to a nondegenerate and one to a degenerate species: The characters of the product are the products of the individual characters, and represent the characters of some species or irreducible representation. Thus, in C_{4v} used above as an example with $XeOF_4$, $B_1 \times A_2$ gives the following characters:

$$I: +1 \times +1 = +1; \qquad 2C_4: -1 \times +1 = -1;$$
$$C_4{}^2: +1 \times +1 = +1; \qquad 2\sigma_v: +1 \times -1 = -1;$$
$$2\sigma_d: -1 \times -1 = +1$$

hence $B_1 \times A_2 = B_2$. Also $A_2 \times E$: $+1 \times +2 = +2$; $+1 \times 0 = 0$; $+1 \times -2 = -2$; $-1 \times 0 = 0$; $-1 \times 0 = 0$; $-1 \times 0 = 0$, hence $A_2 \times E = E$.

The problem becomes more difficult when we attempt to form the product $E \times E$: $+2 \times +2 = +4$; $0 \times 0 = 0$; $-2 \times -2 = +4$; $0 \times 0 = 0$; $0 \times 0 = 0$. Obviously, these are not the characters of any *one* species; such a set of characters is called a *reducible representation*, which can in general be reduced to the sum (called the *direct sum*) of several irreducible representations.

In such a direct sum, the character of each operation in the reducible representation is the sum of the characters of the component irreducible representations. In the majority of cases the analysis may be made by inspection. Obviously, the character $+4$, of I can arise only by: (1) $E + E$. (2) $E + 2\Gamma_{ND}$ (Γ_{ND} is either of the nondegenerate A or B, or any combination of them) or (3) $4\Gamma_{ND}$ (in any combination). The character 0 of σ_v indicates as many A_1 plus B_1 as A_2 plus B_2, the character 0 for σ_d as many A_1 plus B_2 as A_2 plus B_1. The character 0 of C_4 demonstrates as many A as B, and finally, the $+4$ of $C_4{}^2$ shows that there are four A's and B's, and hence case 3 holds. Combination of all this information shows that the direct sum is $A_1 + A_2 + B_1 + B_2$.

What is the physical meaning of this finding? In the preceding work, we found that multiplication of any *two* functions of given symmetry gave *one* new function of some symmetry determined by the character multiplication. But here we find that multiplication of *two degenerate* functions gives something new that may be described either as one new function with *four* components of differing symmetry, or as four separate functions of different symmetries.

This matter of forming the product of functions of symmetry species is perfectly general. The rule of multiplying the characters arises from the treatment in terms of the transformation matrices given in Chapter 4. Any doubly degenerate function E can in general be expressed as a linear combination of two functions, $E_1 + E_2$. Similarly, another degenerate function $E' = E_1' + E_2'$. Multiplication of E by E' gives $E \times E' = E_1 E_1' + E_1 E_2' + E_2 E_1' + E_2 E_2'$, a linear combination of four products. Consequently, it is to be expected that the transformation matrix of the product $E \times E'$ is a 4×4 matrix. Such a matrix is the *direct product* of the two individual matrices, formed by the rule:

$$\begin{pmatrix} a_{11} & a_{12} \\ a_{21} & a_{22} \end{pmatrix} \times \begin{pmatrix} b_{11} & b_{12} \\ b_{21} & b_{22} \end{pmatrix} = \begin{pmatrix} a_{11}b_{11} & a_{11}b_{12} & a_{12}b_{11} & a_{12}b_{12} \\ a_{11}b_{21} & a_{11}b_{22} & a_{12}b_{21} & a_{12}b_{22} \\ a_{21}b_{11} & a_{21}b_{12} & a_{22}b_{11} & a_{22}b_{12} \\ a_{21}b_{21} & a_{21}b_{22} & a_{22}b_{21} & a_{22}b_{22} \end{pmatrix}$$

For the case of E^2 of C_{4v}, for the C_4 operation, which was discussed in detail in Chapter 4, this gives:

$$\begin{pmatrix} 0 & -1 \\ 1 & 0 \end{pmatrix} \times \begin{pmatrix} 0 & -1 \\ 1 & 0 \end{pmatrix} = \begin{pmatrix} 0 & 0 & 0 & 1 \\ 0 & 0 & -1 & 0 \\ 0 & -1 & 0 & 0 \\ 1 & 0 & 0 & 0 \end{pmatrix}$$

The character of the direct product is $a_{11}b_{11} + a_{11}b_{22} + a_{22}b_{11} + a_{22}b_{22} = (a_{11} + a_{22}) \cdot (b_{11} + b_{22})$, or in other words, the product of the characters of the two individual transformation matrices, in the example equal to zero.

The same type of procedure applied to the nondegenerate, or to one degenerate and one nondegenerate function shows that the result stated above follows immediately from the direct product. The product of two nondegenerate functions is a single function, the direct product of two 1×1 matrices (a_{11}) and (b_{11}) is just the 1×1 matrix $(a_{11}b_{11})$, which is identical to its character. For $A \times E = A \times (E_1 + E_2)$, the transformation matrices $a_{11} \times \begin{pmatrix} b_{11} & b_{12} \\ b_{21} & b_{22} \end{pmatrix}$ give $\begin{pmatrix} a_{11}b_{11} & a_{11}b_{12} \\ a_{11}b_{21} & a_{11}b_{22} \end{pmatrix}$ with character $a_{11}b_{11} + a_{11}b_{22} = a_{11}(b_{11} + b_{22})$, again the product of the characters. The same direct product method can be applied to species of higher degeneracy, and gives again the same result.

We have shown how the reducible representation can be reduced to a direct sum of irreducible representation by inspection in some relatively simple cases. How can this be achieved for cases where the inspection method fails to give an answer with reasonable effort? Without proof we state that the number n_Γ of functions in the product belonging to the irreducible representation Γ is given by:

$$n_\Gamma = \frac{1}{g} \sum g_R \gamma_\Gamma{}^R \gamma_{DP}{}^R \tag{5.6}$$

where g is the order of the group, g_R the order of the class to which the symmetry operation R belongs, $\gamma_\Gamma{}^R$ its character in the species Γ, and $\gamma_{DP}{}^R$ its character in the direct product, and the summation extends over all classes of operations.

Take as an example the point group C_{4v}. Its order is the total number of operations, or the sum of the squares of the degeneracies of its irreducible representation, in this case 8: either

$$1I + 2C_4 + C_4{}^2 + 2\sigma_v + 2\sigma_d = 8$$

or

$$1^2(A_1) + 1^2(A_2) + 1^2(B_1) + 1^2(B_2) + 2^2(E) = 8.$$

The characters of $E \times E$ are 4, 0, 4, 0, 0. Then eq. (5.6) gives for $E \times E$:

$$n_{A_1} = \tfrac{1}{8}(1 \times 1 \times 4 + 2 \times 1 \times 0 + 1 \times 1 \times 4 + 2 \times 1 \times 0 + 2 \times 1 \times 0)$$
$$= \tfrac{8}{8} = 1$$

$$n_{A_2} = \tfrac{1}{8}(1 \times 1 \times 4 + 2 \times 1 \times 0 + 1 \times 1 \times 4 - 2 \times 1 \times 0 - 2 \times 1 \times 0)$$
$$= \tfrac{8}{8} = 1$$

$$n_{B_1} = \tfrac{1}{8}(1 \times 1 \times 4 - 2 \times 1 \times 0 + 1 \times 1 \times 4 + 2 \times 1 \times 0 - 2 \times 1 \times 0)$$
$$= \tfrac{8}{8} = 1$$

$$n_{B_2} = \tfrac{1}{8}(1 \times 1 \times 4 - 2 \times 1 \times 0 + 1 \times 1 \times 4 - 2 \times 1 \times 0 + 2 \times 1 \times 0)$$
$$= \tfrac{8}{8} = 1$$

$$n_E = \tfrac{1}{8}(1 \times 2 \times 4 + 2 \times 0 \times 0 - 1 \times 2 \times 4 + 2 \times 0 \times 0 + 2 \times 0 \times 0)$$
$$= \tfrac{0}{8} = 0$$

and $E \times E = A_1 + A_2 + B_1 + B_2$, as seen above. This method can of course readily be applied to a direct sum of any complexity.

In most cases, in the direct product of several degenerate vibrations, there occur some degenerate vibrations. In C_{3v}, for example, $E \times E$ gives $A_1 + A_2 + E$, apparently only three functions; since E, however, represents the sum of two, we still have four functions.

Having thus determined how to multiply degenerate functions, let us return to the problem of selection rules and polarization. Let us take the point group D_{4d} as an example and work out the electronic transitions that are allowed and those that are forbidden. The transitions $a_1 \leftrightarrow a_1$, $a_2 \leftrightarrow a_2$, $b_1 \leftrightarrow b_1$, and $b_2 \leftrightarrow b_2$ give $\psi_i \psi_j$ products belonging to A_1, $a_1 \leftrightarrow a_2$, $b_1 \leftrightarrow b_2$, belonging to A_2, $a_1 \leftrightarrow b_1$, $a_2 \leftrightarrow b_2$, to B_1, $a_1 \leftrightarrow b_2$, $a_2 \leftrightarrow b_1$ to B_2; $a_1 \leftrightarrow e_1$, $a_2 \leftrightarrow e_1$, $b_1 \leftrightarrow e_3$, and $b_2 \leftrightarrow e_3$ to E_1; a_1, a_2, b_1, $b_2 \leftrightarrow e_2$ to E_2; a_1, $a_2 \leftrightarrow e_3$, b_1, $b_2 \leftrightarrow e_1$ to E_3. The transitions $e_1 \leftrightarrow e_2$ and $e_2 \leftrightarrow e_3$ have two components; $E_1 + E_3$, $e_1 \leftrightarrow e_1$, and $e_3 \leftrightarrow e_3$, three, $A_1 + A_2 + E_2$ and $e_1 \leftrightarrow e_3$, three,

$B_1 + B_2 + E_2$ and $e_2 \leftrightarrow e_2$, four, $A_1 + A_2 + B_1 + B_2$. Of the products having only one component, those with B_2 and E_1 are allowed, and z and xy polarized, respectively. Thus $a_1 \leftrightarrow b_2$ and $a_2 \leftrightarrow b_1$ are allowed and z, $a_1 \leftrightarrow e_1$, $a_2 \leftrightarrow e_1$, $b_1 \leftrightarrow e_3$, and $b_2 \leftrightarrow e_3$ allowed and xy polarized. Of the transitions between e levels, $e_1 \leftrightarrow e_2$ and $e_2 \leftrightarrow e_3$ are allowed and xy polarized because of the E_1 component in the product, $e_1 \leftrightarrow e_3$ and $e_2 \leftrightarrow e_2$ are allowed and z polarized because of B_2. No transition is nonpolarized in this point group. Thus, of a total of 49 types of transitions, only 16 are allowed, 8 of each type of polarization. We may note in passing that the degeneracy of x and y implies polarization in that plane, but no distinction of these degenerate axes is possible.

The method used here for the determination of selection rules is based on the assumption that we can factor the complete electronic wave function into a product of one-electron wave functions. In this case of degenerate wave functions, however, this procedure is not completely satisfactory. Such products can, and often do, give rise to various states. In the ground states of most stable molecules, where all orbitals are completely occupied, the state is totally symmetric. In excited states, however, where shells of orbitals are incompletely filled, many states of various symmetry exist. These states cannot be completely predicted on the basis of the direct product alone, since the electron spins and the Pauli principle produce additional restrictions on the one hand, but additional states of differing multiplicity on the other. The direct product however, does determine the possible symmetry species to which the various states could belong. In addition, given the symmetry of the ground and excited states of a given transition, the direct product determines the selection rule, as indicated above.

If we attempt to make the same determination for the infrared vibrations of a D_{4d} molecule, we find first that, among the non-degenerate species, only b_2 vibrations are active (and z polarized), since here the $\Delta v = 1$ transition requires $a_1 \rightarrow b_2$ or $b_2 \rightarrow a_1$ transitions. Similarly for $v = 0 \rightarrow v = 1$ transitions, only e_1 vibrations are active and xy polarized. Among combination bands, in addition, the combinations $a_1 b_2$ and $a_2 b_1$ are active and z polarized, and $a_1 e_1$, $a_2 e_1$, $b_1 b_3$, and $b_2 e_3$ are active and xy polarized if both transitions are $v = 0$ to $v = 1$.

To determine the activity of higher jumps ($\Delta v = 1$ from higher v)

and overtones ($\Delta v > 1$), we must first gain information on the symmetry species of the higher wave functions of degenerate vibrations. In a doubly degenerate vibration, the quantum number v is the sum of two components, which may be considered as the separate quantum numbers of the two orthogonal components of the vibration. For $v = 1$ we have two possibilities, $v_x = 0$, $v_y = 1$, or $v_x = 1$, $v_y = 0$, and hence a doubly degenerate wave function. For $v = 2$ we have three possibilities: 2,0; 1,1; or 0,2; and hence

(a)   (b)

Fig. 5.10   All *cis*- (a) and all *trans*-octachlorocyclooctane (b).

a triply degenerate function, and in this way the degeneracy keeps increasing. Obviously, a triply degenerate function cannot belong to a doubly degenerate species, nor be composed of two components each belonging to a doubly degenerate species. The determination of the combination of species to which these functions belong is a complicated matter, and we will be satisfied by giving the results here for D_{4d}, and in Appendix 3 for other point groups. We will give the v value as a superscript on the species, followed by the direct sum: $(e_1)^2$, $A_1 + E_2$; $(e_1)^3$, $E_1 + E_3$; $(e_1)^4$, $A_1 + B_1 + B_2 + E_2$; $(e_2)^2$, $A_1 + B_1 + B_2$; $(e_2)^3$, $2E_2$; $(e_2)^4$, $2A_1 + A_2 + B_1 + B_2$; the e_3 behave exactly like e_1. From this it is readily shown that the overtones $(e_1)^0 \rightarrow (e_1)^3$ are active and xy polarized, and $(e_2)^0 \rightarrow (e_2)^2$, $(e_2)^0 \rightarrow (e_2)^4$, and $(e_1)^0 \rightarrow (e_1)^4$ are active and z polarized. It is also apparent that $(\Gamma)^2$ always has a component A_1 (or in other point groups a totally symmetric component), and hence the $v = 1 \rightarrow v = 2$ transition of all allowed vibrations is allowed. The same holds for higher vibrations.[1]

Let us now apply these considerations to a hypothetical structure determination. Say we have prepared two octachlorocyclöoctanes, but do not know which is the all *cis* and which the all *trans*. Assuming these two have the structures shown in Fig. 5.10, they belong to

---

[1] For a more rigorous treatment of overtone and combination bands see the references at the end of the chapter.

the point groups C_{4v} and D_{4d}, respectively. Using the formulas developed in a preceding section and given in Appendix 2, we find the number of fundamentals shown in Table 5.7. For the allowed vibrations the polarization is given. Thus the *cis* compound has 27, the *trans* compound only 13 infrared-active fundamentals.

Table 5.7   The Number of Fundamentals of *cis*- and *trans*-Octachloro-cyclooctane

Species	No. $cis(C_{4v})$	Polar.	Species	No. $trans(D_{4d})$	Polar.
A_1	11	z	A_1	6	
A_2	5		A_2	2	
B_1	6		B_1	3	
B_2	12		B_2	5	z
E	16	x, y	E_1	8	x, y
			E_2	9	
			E_3	8	

Unfortunately, not all of these are readily resolved, since, of the eight CH and the eight CCl stretching vibrations in the *cis* compound, three each will be active, and very likely will be difficult if not impossible to resolve. Still, the *cis* compound will have many more fundamentals showing up.

If we can make some very high resolution measurements and observe a number of overtones and combination bands, we should be able to obtain further information. In the *trans* compound, we have just seen that a_1b_2, a_2b_1, a_1e_1, a_2e_1, b_1e_3, and b_2e_3 combinations are active, the others inactive. In the *cis* compound, it is easy to verify that the combinations a_1a_1, a_2a_2, b_1b_1, b_2b_2, a_1e, b_1e, and b_2e are active. In the *trans* compound, *no* combinations of two active frequencies occur, but in the *cis* compounds, the *active* combinations a_1a_1 and a_1e are combinations of *active* fundamentals. In the overtones, also, there is a distinct difference. In the *trans* compound, certain overtones of the various e's occur; otherwise, only those of b_2. In the *cis* compound, all *even* overtones of the nondegenerate species are active, and for the degenerate species all overtones, even and odd. These differences should suffice to distinguish the molecules and possibly even to identify one in the absence of the other.

Raman Spectra.   The treatment of selection rules in Raman spectra is quite similar to that for infrared spectra. Raman spectra arise when molecules are irradiated with light that is not absorbed but scattered. Some of the scattered light has a different frequency, and this frequency difference corresponds to the energy of vibrational and rotational transitions. Considering only the vibrational Raman spectra, the vibrational wave functions are of course the same as those discussed above. The intensity of a Raman line is determined by an integral quite analogous to that of eq. (5.5b) except that the dipole moment **M** vector is replaced by the induced dipole moment vector **P**. The dipole moment induced in the molecule by the exciting radiation (that is, the radiation being scattered) is given by

$$\mathbf{P} = \alpha\mathbf{E}$$

where **E** is the electric vector of the incident radiation and α is called the polarizability tensor or polarizability matrix, a 3×3 matrix with components α_{xx}, α_{xy}, α_{xz}, α_{yx}, etc. The intensity-determining integral

$$\int \chi_i \mathbf{P} \chi_f \, d\tau$$

then has nine components. However, the matrix α is symmetric about its principal diagonal so that $\alpha_{xy} = \alpha_{yx}$, etc.; hence only six of the components are distinct. If the integrand of any one of the six component integrals is totally symmetric, this component is nonvanishing, and the total integral is *not* zero. In this case the vibration is said to be *Raman active*.

It is not an easy matter to determine the symmetry species to which a given polarizability element α belongs, and the methods and proofs will not be given. The components are given in the character tables, however, together with the species of translations and rotations, in Appendix 1.

In point group D_{4d}, α_{xx} and α_{yy} transform together (that is, in linear combinations) like A_1 and E_2, α_{zz} transforms like A_1, α_{xy} transforms like E_2, and α_{xz} and α_{yz} transform like E_3. Thus vibrations a_1, e_2, and e_3 are Raman active. In *trans*-octachlorocyclooctane, there are thus 23 Raman-active fundamentals, none of which are infrared active. In C_{4v}, α_{xx} and α_{yy} transform jointly like A_1 and B_1, α_{zz} like A_1, α_{xy} like B_2, and α_{xz} and α_{yz} like E. Hence all fundamentals except a_2 are Raman active—a total of 45 for *cis*-octachlorocyclooctane; again, a very much larger number. Moreover, all infrared-active fundamentals are also Raman active, and observation of both spectra makes the distinction quite easy.

Just as in the infrared molecules with a center of symmetry have a somewhat special status, so also in Raman. Where all components of M are ungerade, all components of α are gerade, and only gerade vibrations can be Raman active. Since only ungerade vibrations are infrared active, coincidences between Raman and infrared lines immediately eliminate the possibility of a center of symmetry, unless the coincidence is due to an accidental degeneracy, that is, two fundamentals of different symmetry having the same frequency.

One additional interesting feature of Raman spectra is the fact that the scattered light usually is polarized. The degree of polarization (or depolarization), however, varies depending on the symmetry of the polarizability matrix and on the symmetry of the excited vibration. In particular, Raman lines of totally symmetric fundamentals are polarized, and those of other fundamentals are depolarized. Thus, measurement of the polarization of the Raman line provides a means of determining experimentally which fundamentals are totally symmetric.

Finally, it may be worthwhile to mention just briefly the selection rules in microwave spectra. The requirement for a microwave spectrum to be observed is the presence in the molecule of a permanent dipole moment. This eliminates immediately from consideration all spherical top molecules, that is, all molecules with more than one axis of order higher than two, and all molecules with a center of symmetry. Symmetric tops, that is, molecules with one but only one axis above twofold, have relatively simple microwave spectra because the dipole moment coincides with the axis.

REFERENCES

Gerhard Herzberg, *Molecular Spectra and Molecular Structure*, Vol. 2, *Infrared and Raman Spectra of Polyatomic Molecules*, D. Van Nostrand Co., Princeton, N.J., 1945, Chapter 2, Sect. 4.

E. B. Wilson, Jr., J. C. Decius, and P. C. Cross, *Molecular Vibrations*, McGraw-Hill Book Co., New York, 1955, Chapters 6 and 7.

H. H. Jaffé and Milton Orchin, *Theory and Applications of Ultraviolet Spectroscopy*, John Wiley and Sons, New York, 1962, Chapter 6.

H. B. Gray and C. J. Ballhausen, *J. Am. Chem. Soc.*, **85**, 260 (1963).

6

Crystal Symmetry[1]

....

6.1 THE SOLID STATE

One of the most important areas of application of principles of symmetry in chemistry is the description, and hence the determination, of the structure of crystalline solids. A crystalline solid may be defined as a material made up of an infinitely (or practically infinitely) repeating arrangement of matter. This problem of the repetition introduces some new aspects into the symmetry con-

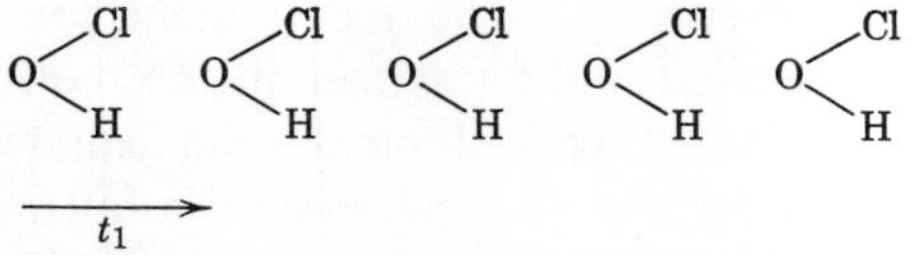

Fig. 6.1 One-dimensional translation, t_1.

siderations; in preceding chapters we have been concerned with symmetry operations which leave the center of gravity of the system fixed, and hence with the point group. To these operations we must now add the operations needed in the treatment of a solid, namely, those which repeat the basic unit throughout the crystal. Such an operation is called a *translation*; it is most easily illustrated in a single dimension. Take some arbitrary molecule, say HOCl, and repeat it as many times as you wish after moving a given distance

[1] In agreement with the custom of crystallographers, we use International notation for symmetry operations, point groups and space groups in this chapter. This notation uses a number (p) for a p-fold axis, an m for a plane, and a barred number $\bar{p}$ for a p-fold rotation-inversion axis.

along a straight line (Fig. 6.1). This represents the operation of translation in one dimension, t_1. Similarly, we can define translation in two dimensions by translating the whole of Fig. 6.1 any number of times in a direction other than that of t_1 (cf. Fig. 6.2). If we now replace the center of gravity of each molecule (or the 0 atom, or

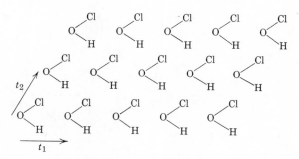

Fig. 6.2   Two-dimensional translation, t_1 and t_2.

any point) by a dot, we obtain what is called a *lattice*, as shown in Fig. 6.3.

We can now choose any one of a number of areas of Fig. 6.2 or 6.3, and consider this area as the unit of repetition, called the *unit cell*; Fig. 6.4 shows a few choices. At this point the choice is

Fig. 6.3   A two-dimensional lattice.

arbitrary. The unit cells a, b, and c of Fig. 6.4 each contain only a single point, which can be seen in either of two ways, both shown in Fig. 6.5.

(*a*) If the unit cell is moved along each translation by some fraction of the distance between two points, it is seen that each unit cell contains just a single point; or (*b*) the point at each corner of the unit cell belongs equally to four unit cells; since the cell has four corners, one-fourth of each point belongs to each cell: $4 \times \frac{1}{4} = 1$.

Since the unit cells a, b, and c of Fig. 6.4 contain only one point each, they are called *primitive*. Cells like d are called multiple;

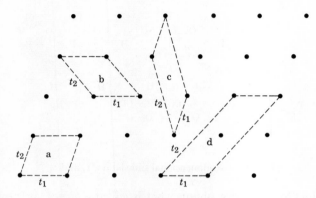

Fig. 6.4   The unit cell.

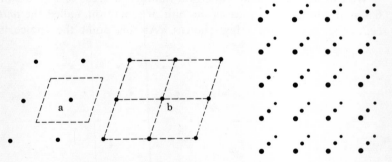

Fig. 6.5   The primitive unit cell.   Fig. 6.6   A three-dimensional lattice.

double if they contain two, as d does, or triple, and so on. The two translations t_1 and t_2, along which the lattice repeats, are called *conjugate* translations. The choice of a unit cell is determined by convenience and convention to be the cell that best represents the symmetry of the lattice.

A lattice can, of course, be extended into three dimensions by a third translation, t_3, which is not coplanar with the other two, as shown in Fig. 6.6. Similarly, an *array* or *lattice array* of real objects repeated by translations, distinguished from a lattice of imaginary points, can be extended into a third dimension.

6.2 PLANE LATTICES AND PLANE GROUPS

Next it is interesting to inquire about the symmetry properties permitted by the unit cell of a planar lattice, that is, a lattice in two-dimensional space. The molecule so far considered, HOCl, has no element of symmetry in the two-dimensional space defined by its plane. This absence of any element of symmetry within the plane permits a perfectly general lattice, with different unit lengths for t_1 and t_2, called a and b, and any angle γ between them (Fig. 6.7a). The same lattice, called a *parallelogram* lattice, can accommodate a two-dimensional molecule with a center of symmetry, for example, *trans*-dichloroethylene accommodated in a parallelogram lattice Fig. 6.8.

In order better to correlate two- and three-dimensional symmetry, it is convenient to place the plane lattice formed by the application of two translational operations into three-dimensional space.[2] Then the center of symmetry just discussed is equivalent to a twofold rotational axis normal to the plane defined by the two translations. Similarly, a line of symmetry, as in the two-dimensional H_2O molecule, becomes the plane of symmetry perpendicular to the molecular plane.

A plane of symmetry (m) requires that the lattice points lie on rows both parallel and perpendicular to m, resulting in either a *rectangular* (Fig. 6.7b, $\gamma = 90°$) or a *diamond* lattice (Fig. 6.7c_1, $a = b$ or $a_1 = a_2$). Both these lattices also accommodate a second plane and a twofold axis. In the case of the diamond lattice, it is conventional to replace the diamond-shaped, primitive unit cell by a double rectangular one, as shown in Fig. 6.7c_2, which contains a lattice point at its center, and hence is generally called c in contrast to the

---

[2] Then, of all possible symmetry operations, only those that do not move atoms lying in the lattice plane out of this plane are allowed.

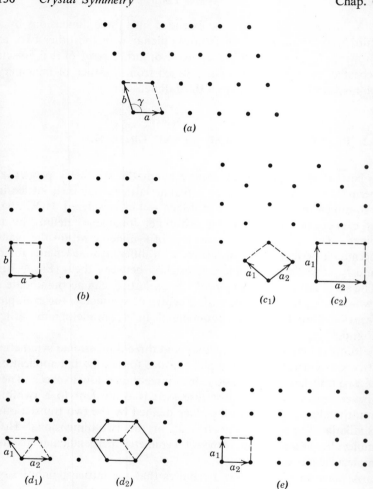

Fig. 6.7   The five kinds of plane lattices.

primitive cell called p. A special case of the diamond lattice in which the angle between a_1 and a_2 is 60 or 120° is shown in Fig. 6.7d_1. It is called the *rhombus* or *triangular* lattice, since two edges and a short diagonal enclose an equilateral triangle. This lattice can accommodate a three- or a sixfold axis, aside from the three planes; the sixfold axis is best seen in terms of the *hexagonal* lattice (a benzene or graphite lattice) shown in Fig. 6.7d_2. The unit cell best suited for a sixfold axis, as shown in Fig. $6.7d_2$, is a hexagonal cell, which is three times the minimum unit cell which is a rhombus. Finally, the most symmetric lattice is a *square* lattice with $a_1 = a_2$, and enclosing a right angle γ (Fig. 6.7e).

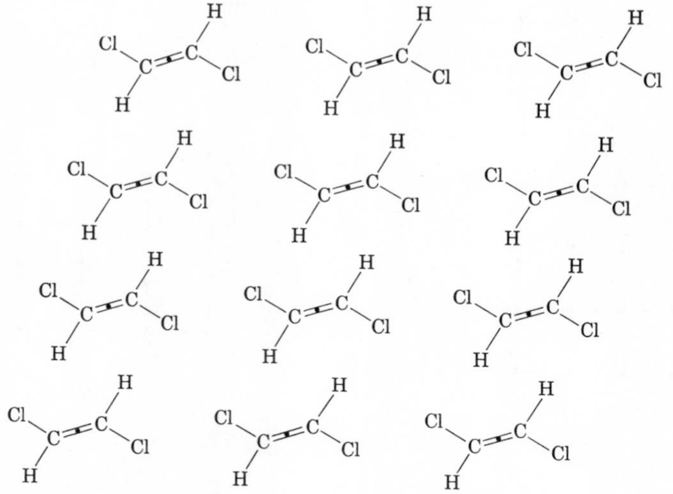

Fig. 6.8   A parallelogram lattice of a molecule with a center of symmetry.

The five lattices of Fig. 6.7 are the only possible ones in two dimensions. Since a repetition of the unit cell must completely occupy all space, it is readily seen from Fig. 6.9 that any one of the unit cells of these five lattices fulfills this condition; no other figures can be found which also fulfill this requirement. This implies that the only elements of symmetry allowed in the unit cell of a planar array are planes, *m*, and axes, *2, 3, 4,* and *6*. *Molecules* with other elements of symmetry must occur more than once in a unit cell, in such a manner that the elements of symmetry of the *cell* are restricted to the above.

The restriction of the rotational axes to those cited may be seen in a different way—consider Fig. 6.10. Take one row of a plane lattice, and rotate about a *p*-fold axis by the angle $\varphi = 2\pi/p$ at two adjacent points. Since *p* rotations by φ superimpose on the starting configuration, it does not matter that we rotate clockwise about one and counterclockwise about the other axis. The two new lattice

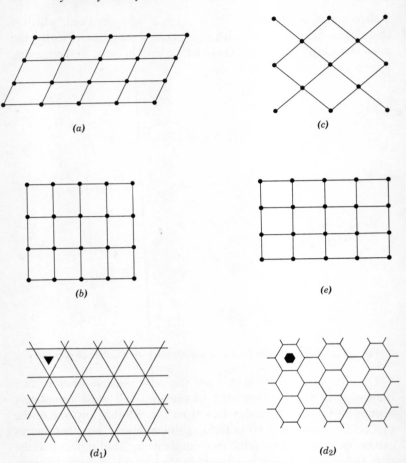

Fig. 6.9   Filling a plane by unit cells. Note that (d_1) and (d_2) are equivalent.

points produced by these rotations occur at k and l; these points are equidistant from the original row by construction. Hence the line joining k and l is parallel to the translation t; their distance apart must be some integral multiple m of the unit translation t, that is, mt, otherwise the line kl is not a translation, and the array resulting from the rotation by φ is not periodic. The distance kl is $mt = t + 2t \cos \varphi$, and for $m = 0, \pm 1, \pm 2, \ldots$. This gives $\cos \varphi = (m - 1)/2$ or $\cos \varphi = N/2$, where the integer N equals $m - 1$.

Values of N outside the range -2 to $+2$ give no values of φ, since $\cos \varphi$ lies between -1 and $+1$, inclusive. The five possible values of N give the following values of φ: -2, $180°$; -1, $120°$; 0, $90°$; $+1$, $60°$; and $+2$, $0°$ or $360°$. Hence the axes may be two-, three-, four-, six-, or onefold, respectively, and no other axes are possible.

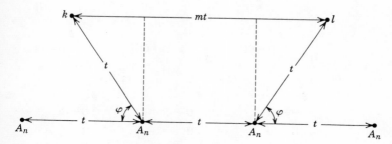

Fig. 6.10   A rotational axis in a plane lattice.

We can now combine the translation operation with the symmetry operation of our planar unit cell. What happens when we combine a translation in a parallelogram lattice with a twofold (*2*) axis at the lattice point? To examine this, we use a primitive unit cell with two HOCl molecules symmetrically placed with respect to the lattice point, as in Fig. 6.11*a*. Translation in the two directions multiplies these twofold axes as well as the molecules, as shown in Fig. 6.11*b*. It is readily seen by inspection of Fig. 6.11*b* that new twofold axes immediately arise, indicated by crosses in Fig. 6.11*b*, which are shown in Fig. 6.11*c*, one along each translation line, halfway between each pair of lattice points, and another one halfway between lattice points on the unit cell diagonal.

In a similar manner, combination of a threefold axis with a translation produces a new threefold axis, and the combination of a *4* with *t* produces a new *4* and new *2*'s. Combination of a *6* with *t* gives two *3*'s and a series of *2*'s; all these relations are illustrated in Fig. 6.12. Such combinations of symmetry operations with plane lattices are called *plane groups*. Figure 6.12 shows all the plane groups formed from rotational axes and translations alone. Their designations are shown using a *p* to indicate the *p*rimitive unit cell, combined with the notation for the highest rotational axis, from

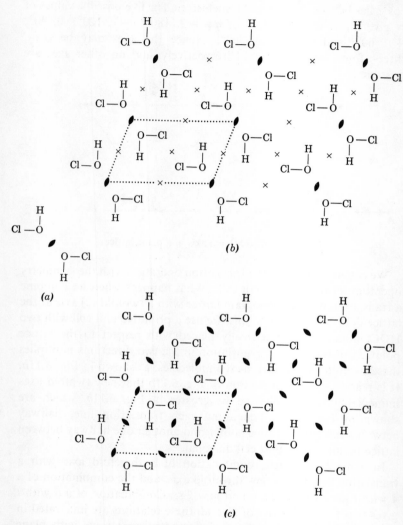

Fig. 6.11   Generation of new twofold axes by translation and rotation.

which the group was constructed. The lowest plane group, for which there is no element (that is, a *1* at the lattice point), produces of course no new elements on translation and is called *p1*.

A single plane of symmetry, parallel to t_1 in a rectangular lattice, on translation produces a new plane halfway between two planes and gives rise to the plane group *pm*, illustrated in Fig. 6.13. Two such

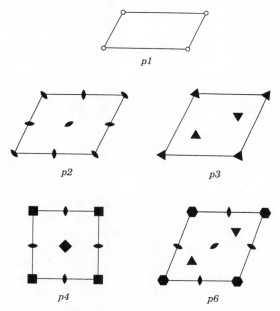

Fig. 6.12   The plane groups *p1, p2, p3, p4*, and *p6*.

planes at right angles to one another imply a twofold axis at the lattice point, and give rise on translation to two new planes, halfway between the old ones, and to the new *2*'s as in *p2*; this is plane group *pmm* (or *p2mm*), shown in Fig. 6.14; the *2* is not necessary in the designation, since it is implied by the *mm*. The *p* specifies a primitive unit cell since it is seen in Fig. 6.13 that only one molecule occurs per unit cell.

In order to discuss the remaining possible plane groups, we must introduce a new symmetry operation. This is the combination of a reflection at a plane coupled with a translation by one-half the translational unit, and is called a *glide plane* (actually, in three dimensions; in two dimensions a *glide line* is more accurate, as in two dimensions any plane of symmetry ought to be called a *line*). The simplest way of illustrating a glide plane may be to examine a case in which one arises by itself.

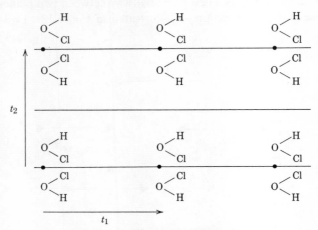

Fig. 6.13   The plane group *pm*.

Examine a simple plane (*m*) in a diamond lattice; the unit cell in this lattice is conventionally chosen not as the primitive rhombus, but as a centered rectangle (Fig. 6.7c_2). It is designated by *c*. If a pair of HOCl molecules in such a lattice is reproduced by translation, Fig. 6.15 results, where the solid lines are the mirror planes.

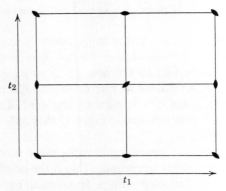

Fig. 6.14   A single unit cell in plane group *pmm*.

Here a new type of symmetry has arisen. Translation by one-half of a translational unit along t_1 and reflection on the dashed plane transports each molecule into the position of another one, and hence is a symmetry operation, a glide plane. It is indicated graphically by a dashed line. Note that this element is similar in some respects to a rotation-reflection or a rotation-inversion axis, in that it is a combination of two successive operations, neither of which, by

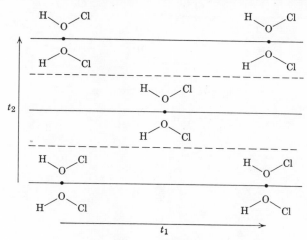

Fig. 6.15   The glide plane in *cm*.

itself, is a symmetry operation in this particular case. In the plane group just described, the glide plane arose as a consequence of the other operation, *m*; the group is called *cm*, since *m* alone is sufficient to define the whole symmetry in the lattice *c*, or alternately, *cg* (*g* for glide plane).

Glide planes can exist alone. A single glide plane in a rectangular lattice, coupled with translation, gives rise to the plane group *pg* (Fig. 6.16).

We have thus briefly described nine plane groups: *p1*, *p2*, *p3*, *p4*, *p6*, *pm*, *pmm*, *cm*, and *pg*. There is a total of seventeen such groups. The remainder are quite complicated and may be found in the International Tables of Crystallography and many other places.

Finally, let us examine the relation between molecular symmetry and crystal symmetry. The HOCl molecule, which we have repeatedly used as an example, has no elements of symmetry in the

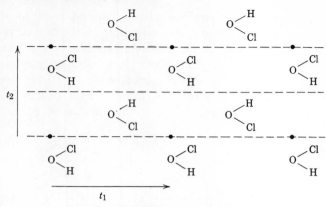

Fig. 6.16   The plane group *pg*.

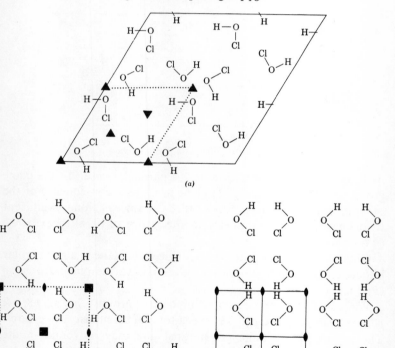

Fig. 6.17   Various lattices of HOCl (*a*) *p3*, (*b*) *p4*, (*c*) *pmm*.

plane. Can we hence infer that it must crystallize—still in a hypothetical plane crystal—in a lattice without symmetry, that is, in the
plane group *p1* shown in Fig. 6.12? We have already seen that the
answer must be no. Figure 6.11 showed an arrangement of these
molecules in a *p2* lattice, Fig. 6.13 in *pm*, and Fig. 6.15 in *cm*. Thus,
by placing two molecules in the unit cell, we have been able to

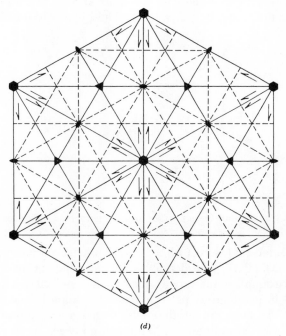

(d)

Fig. 6.17(*d*)   *p6m*.

obtain a *2*, an *m*, or an *m* and a *g*, in the unit cell, even though no *2*
or *m* or *g* is present in the molecule.   Inclusion of three molecules
permits the generation of a unit cell with a *3* (*p3*) (Fig. 6.17*a*); four
molecules can lead to several possibilities, including a *4* (*p4*) or a *2*
and two *m* (*pmm*) (Fig. 6.17*b* and *c*); and twelve molecules may even
give *p6m*, shown in Fig. 6.17*d*. Thus inclusion of enough molecules,
appropriately placed, permits construction of a unit cell of any
desired symmetry, belonging to any desired plane group.   As a
consequence, we can find in nature some completely unsymmetric

molecules crystallized in crystals of high symmetry, whereas much more symmetric crystals frequently crystallize with much lower symmetry.

The reverse of this relation is equally true: A molecule of high symmetry can readily crystallize in a lattice in which the unit cell has much lower symmetry. That this must be may be seen from a consideration of molecules with rotational axes other than those permitted in a unit cell (*2, 3, 4,* or *6*). Thus, in a crystal of ferrocene,

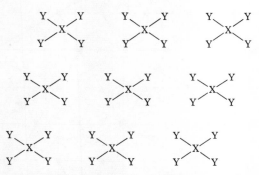

Fig. 6.18   A parallelogram lattice array of tetragonal molecules XY_4.

dicyclopentadienyl iron, with a *5* axis, the unit cell must obviously have a symmetry considerably different, and probably lower. Another example is the way a square planar molecule, XY_4, may be arranged in plane group *p2* with a parallelogram lattice, as shown in Fig. 6.18. Even though the molecule has a *4*, and four *m*'s, the unit cell is based only on *2*'s. Although on *chemical* grounds such a crystal may be unlikely and possibly rare, on *symmetry* grounds it is perfectly possible and permissible.

6.3   SPACE LATTICES AND SPACE GROUPS

It now remains to generalize the discussion of lattices and plane groups of the last section to three dimensions. This discussion is extremely brief because in principle it resembles the discussion of the two-dimensional cases, although in practice it is complicated by the large variety of space groups.

First, it must be realized that the restriction of axes to 1, 2, 3, 4, and 6 demonstrated above for unit cells in two dimensions holds equally in three, since a three-dimensional lattice may be considered formed from a two-dimensional one by repeating the two-dimensional one periodically according to a translation t_3 (c) in the third dimension. However, use of the third dimension also permits rotation-reflection or rotation-inversion axes (in conformance with the custom of crystallographers, we consider only rotation-inversion axes), and again only $\bar{1} = i$, $\bar{2} = m$, $\bar{3}$, $\bar{4}$, and $\bar{6}$ are consistent with the translational symmetry. This limitation restricts the crystallographer to the thirty-two point groups listed in Table 6.1, with the corresponding Schönfliess notation for comparison.

The thirty-two point groups are divided into six separate classes, the *crystal systems*, according to the shape and relative dimensions of the unit cell, which both depend on its symmetry properties. A cell having no element of symmetry or a simple center is subject to no restriction, other than that of filling all space by translational repetition. Such a unit cell is accordingly a parallelepiped, with different edges, $a \neq b \neq c$, and different angles, $\alpha \neq \beta \neq \gamma \neq 90°$ (where α is the angle opposite a), and is the unit cell of the *triclinic* system (triclinic for three angles). Introduction of a twofold axis (rotation or rotation-inversion)[3] restricts the unit cell to a rhombic prism, still $a \neq b \neq c$, $\gamma \neq 90°$, but now $\alpha = \beta = 90°$, and defines the *monoclinic* system (monoclinic for one angle). Three twofold axes (either kind or in combination) restricts the unit cell to a rectangular prism, $a \neq b \neq c$, but $\alpha = \beta = \gamma = 90°$, the *orthorhombic* system. A *single* fourfold axis further introduces the restriction $a = b$, that is, the unit cell becomes a square prism, the *tetragonal* system. Four threefold axes require $a = b = c$, which leads to a cube in the *isometric*, *regular*, or *cubic* system. Finally, a single three- or sixfold axis produces the *hexagonal* system, which is the most complicated one because no single unit cell is particularly convenient. A possible nonprimitive unit cell is the hexagonal prism based on a regular hexagon (Fig. 6.19); alternately, one-third of this cell, a rhombic prism based on an equilateral rhombus with

---

[3] At this point we can appreciate the crystallographers' preference for the use of rotation-inversion to rotation-reflection axes. In the classification scheme here used, a p-fold axis may be either a rotation or a rotation-inversion axis, but *not* a p-fold rotation-reflection!

Table 6.1   The Six Crystal Systems

Point Group Symbol				
S^1	$H\text{-}M^2$	System	Unit Cell	Minimum Symmetry Requirements
C_1	1	Triclinic	$\alpha \neq \beta \neq \gamma \neq 90°$	None
S_2	$\bar{1}$		$a \neq b \neq c$	
C_2	2	Monoclinic	$\alpha = \beta = 90°$	One twofold
C_h	m		$\gamma \neq 90°$	axis or one
C_{2h}	$2/m$		$a \neq b \neq c$	mirror plane
D_2	222	Orthorhombic	$\alpha = \beta = \gamma = 90°$	Any combination
D_{2h}	mmm		$a \neq b \neq c$	of three mutu-
C_{2v}	mm			ally perpendic-
				ular twofold
				axes or mirror
				planes
C_3	3	Hexagonal		
D_3	32		$\alpha = \beta = \gamma \neq 90°$	One threefold
S_6	$\bar{3}$		$a = b = c$	axis or one
D_{3d}	$\bar{3}m$	(Rhombedral		threefold inver-
C_{3v}	$3m$	Division)		sion axis
C_6	6	(Hexagonal	or	
C_{3h}	$\bar{6}$	Division)		
C_{6h}	$6/m$		$\alpha = \beta = 90°$	One sixfold axis
D_6	62		$\gamma = 120°$	or one sixfold
D_{3h}	$\bar{6}2m$		$a = b \neq c$	inversion axis
D_{6h}	$6/mmm$			
C_{6v}	$6mm$			
C_4	4	Tetragonal	$\alpha = \beta = \gamma = 90°$	One fourfold
C_{4h}	$4/m$		$a = b \neq c$	axis or one four-
D_4	42			fold inversion
D_{4h}	$4/mmm$			axis
S_4	$\bar{4}$			
D_{2d}	$\bar{4}2m$			
C_{4v}	$4mm$			

[1] Schönfliess notation.
[2] Hermann-Maugin notation.

Table 6.1 (*Continued*)

Point Group Symbol		System	Unit Cell	Minimum Symmetry Requirements
S^1	H-M^2			
T	23	Cubic	$\alpha = \beta = \gamma = 90°$	Four threefold
O	43		$a = b = c$	axes at $109°28'$
T_h	$m3$			to each other
O_h	$m3m$			
T_d	$\bar{4}3m$			

60° angles, is often convenient. (This rhombus is included in Fig. 6.19 by heavy lines.)

It may be worth pointing out that the axial inequalities given for the triclinic, monoclinic, and orthorhombic systems are *allowed* by symmetry. Equality of axes marked as unequal *may* occur accidentally.

This classification of the point groups into crystal classes reminds us of the classification of the point groups according to the possible degeneracy of their symmetry species. Thus, the point groups included in the isometric crystal class are all those permitting threefold degenerate species (with the exception of the point groups of types I and K, which cannot occur in crystallography because of their high-order axes). All the point groups in the tetragonal and hexagonal systems permit doubly degenerate species, as do all point groups with other axes higher than twofold which do not occur in crystallography. The point groups occurring in the other crystal classes have no axes higher than twofold, and hence no degenerate symmetry species. Molecules belonging to the point groups in the isometric system are spherical tops. Those belonging to the point groups in the tetragonal and hexagonal systems (and all others with higher-order axes) are symmetric tops; all other molecules are asymmetric tops.

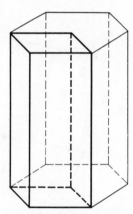

Fig. 6.19   The hexagonal unit cell.

Given thus the unit cells in the six systems, we can proceed to derive the possible space lattices. In the two-dimensional case we encountered primitive (p) and centered (c) lattices. In three dimensions the possibilities are considerably more varied. In each of the systems we can have a primitive lattice, each of which will be denoted by P, and the system to which any one belongs is determined

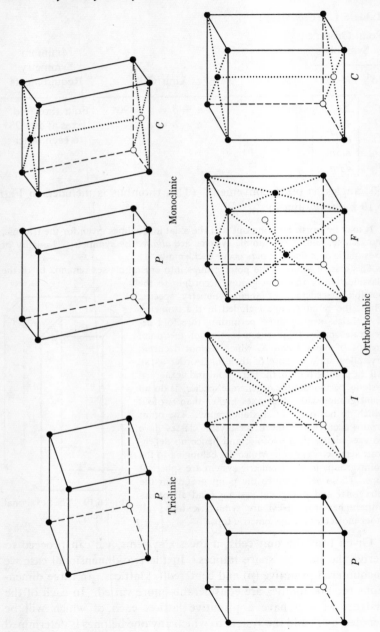

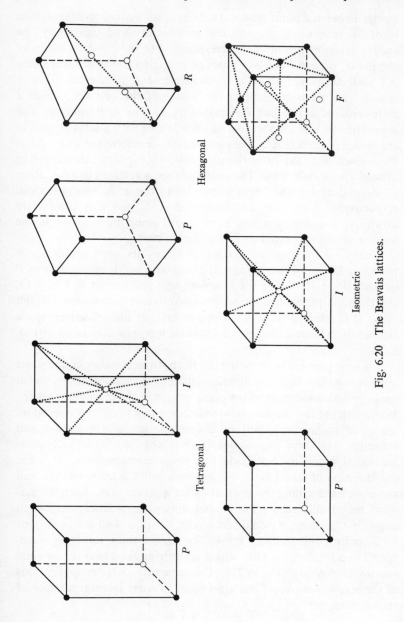

Fig. 6.20 The Bravais lattices.

by the associated point group. In the triclinic system no other lattice is of interest, since the unit cell is arbitrary and can always be readily taken as primitive. Corresponding to the centered lattice of the plane, we can have two types of centering in three dimensions. All systems except triclinic, monoclinic, and hexagonal allow a lattice with a lattice point at the center of the unit cell, called a *body-centered* lattice and designated by *I*. The orthorhombic and isometric systems, in addition, allow a lattice with a lattice point in the center of each face, a *face-centered* lattice referred to as *F*. Also, the monoclinic and orthorhombic systems support face centering of *one* type of face only. The centered face is referred to as *A*, *B*, or *C*, depending on whether it is normal to the *a*, *b*, or *c* direction, respectively. Since the assignment of *a*, *b*, and *c* is completely arbitrary, a convention has been agreed upon by which *b* is the longest of the unit cell axes and *a* the shortest.

Finally, the hexagonal system produces its own problem. A primitive lattice of the hexagonal prism unit cell is referred to as *P* or sometimes *H*; using the alternate unit cell shown in Fig. 6.19, the rhombic prism, a different primitive lattice is possible, though rare. It is referred to as *R*. The unit cells of these fourteen space lattices, which are called the Bravais lattices, are illustrated in Fig. 6.20.

It is now possible to combine the thirty-two crystallographic point groups with the fourteen Bravais lattices into combinations called *space groups*, analogous to the plane groups of the preceding section. We saw in the last section, however, that a new symmetry operation, the glide plane, arose out of a combination of translation and reflection. In three dimensions there is one additional such combinatorial symmetry operation that needs consideration. It is the combination of translation and rotation, called a *screw motion*, and the associated symmetry element called a *screw axis*. Such an axis must be parallel to a translational direction. *p* Rotations by an angle $360°/p$ about a *p*-fold screw axis, coupled with *p* translations by *T*, correspond to just *p* translations, since the *p* rotations correspond to a rotation by $360°$, which is superimposed over the original rotational position (Fig. 6.21). Consequently, after *p* applications of the screw motion, *pT* must be equal to some integral number of translations, *nt*:

$$pT = nt, \qquad T = \frac{nt}{p}$$

Hence the translation accompanying the screw motion must be n/p times the unit translation t. Only values of n/p less than 1 are distinct, being repeated again between each succeeding pair of integers. Thus, for a twofold screw axis n has the unique value of 1, and such an axis is referred to as 2_1. For a threefold screw axis we have two possibilities, 3_1 and 3_2, a threefold rotation coupled with a translation by $\frac{1}{3}$ and $\frac{2}{3}$, respectively, of the unit cell length; these two rotations are mirror images of one another. For 4, we have 4_1,

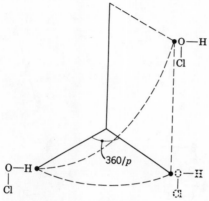

Fig. 6.21 A p-fold screw axis.

4_2, and 4_3, and for 6, we have 6_1, 6_2, 6_3, 6_4, and 6_5. These eleven screw axes can be incorporated with the other symmetry operations of the appropriate point groups.

The combination of the fourteen Bravais lattices with the thirty-two point groups and the *isogonal* symmetry groups, formed by replacing rotational axes by screw axes and mirror planes by glide planes in the point groups, leads to a total of 230 space groups. Of these, 2 are triclinic, 13 monoclinic, 59 orthorhombic, 68 tetragonal, 36 isometric, and 52 hexagonal. It is obviously beyond the scope of this book to derive all of these; a few simple examples must suffice.

In the triclinic system the unique Bravais lattice can combine with either of the only two point groups, 1 or $\bar{1}$, to give the two space groups $P1$ and $P\bar{1}$. In the monoclinic system two Bravais lattices, P or B, can combine with each of three point groups (2, m, or $2/m$), giving a total of six combinations—the space groups *P2*, *B2*, *Pm*, *Bm*, *P2/m*, *B2/m*. In addition, however, the isogonal symmetries give $P2_1$, $B2_1$, *Pb*, *Bb*, $P2_1/m$, $B2_1/m$, *P2/b*, *B2/b*, $P2_1/b$, and $B2_1/b$.

The unit cell of space group *B2* is shown enclosed by a solid line in Fig. 6.22, that of $B2_1$ enclosed by a dashed line. The fact that both unit cells can be shown as alternate subdivisions of the same lattice shows that the two space groups are identical. The same is

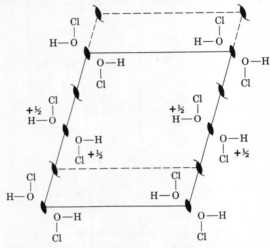

Fig. 6.22   The equivalence of the space groups *B2* and $B2_1$. The face centered molecules, marked with $+1/2$, are one-half the vertical translation above the plane.

true for any pair of space groups of *B* lattices differentiated only by replacement of *2* with 2_1. Consequently, we have to eliminate $B2_1$, $B2_1/m$, and $B2_1/b$ as not independent and are left with fifteen groups, as anticipated.

Notation for the glide planes requires comment. In some systems it is essential to specify their translational directions: *a*, *b*, and *c* specify glide planes parallel to the three edges of the unit cell. The planes *n* and *d* are two special types of glide planes, occurring only in the more symmetrical systems; *n* (for net) is a diagonal one, cutting the unit cell parallel to the diagonal halfway between the diagonal and the corner (cf. Fig. 6.23*a*). The *d* (for diamond) glide plane divides the unit cell into a series of diamonds (cf. Fig. 6.23*b*).

With these comments we can now state the general rules for naming space groups: first the Bravais lattice symbol, followed by the point group symbol, where, for isogonal symmetry, the appropriate replacements are made. Finally, it is interesting to note that for virtually each one of the 230 space groups there exist chemical substances whose crystals belong in that space group.

We have seen above that it is a simple matter in a point group to determine the total number of times a given point must be repeated. This number was largest in any given point group for a general point, one which does not lie on any symmetry element; this number

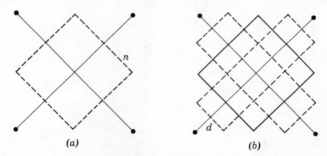

Fig. 6.23   Glide planes (*a*) *n* and (*b*) *d*.

becomes smaller the more symmetry elements a point lies on, to the extent that any point lying on *all* symmetry elements is unique. The same problem recurs in space symmetry, where the number of times an atom is repeated in the unit cell is of great importance. A set of equivalent points in a unit cell is called a set of *equipoints*, and the number of times it occurs is its *rank*.

In this chapter we have, whenever possible, used general points, atoms, or groups of atoms (better atoms in general positions) to illustrate the symmetry relations, and from the diagrams it should be simple to derive the rank of the general equipoints of the various plane groups illustrated. Although generally no atoms have been located on symmetry elements, it should not be difficult to count the rank of any special position. We must only remember that a corner belongs to four cells, and hence all four corners are only one point, and that an edge belongs to two cells, and hence two equivalent atoms lying on edges together represent a single point.

In three dimensions, the problem is quite similar. First, a corner belongs to eight unit cells (except in the hexagonal case, where it belongs to six) an edge to four (except a hexagonal edge to three) and a face to two. If we now examine Fig. 6.22, we find that, as we have drawn it, each of the "corner" HOCl molecules *inside* the unit cell lies in a face. Since there are two molecules in the bottom face of the unit cell which belong jointly to two unit cells, there is *one* molecule (or one of each of the atoms) in the bottom, one in the top face of the unit cell. In addition, there are two more molecules inside the unit cell, in the face center positions, each *completely* in the cell. Hence these atoms, which are general, are of rank 4.

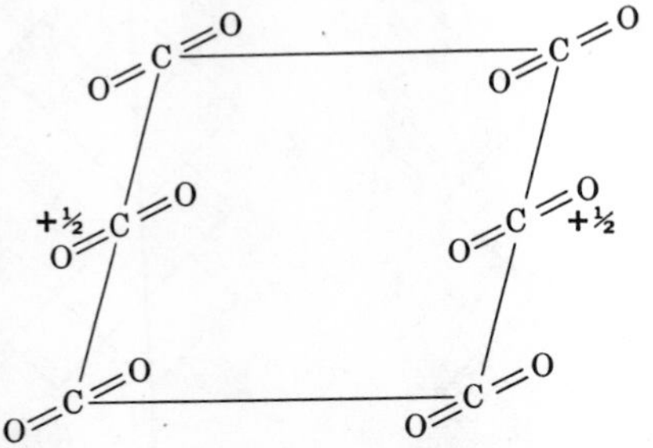

Fig. 6.24   A *B2* lattice of CO_2 molecules.

If we construct a *B2* lattice of CO_2 molecules, placing the C atoms at the lattice points, as in Fig. 6.24, the O atoms are again equipoints of rank 4. The C atoms occurring at each corner define one point and the other two at face centers another one, and we have an equipoint of rank 2.

6.4   CLOSE PACKING

One of the most interesting arrangements of atoms in a crystal is the one called *close packing*, in which the maximum number of atoms is packed into the smallest possible space. This arrangement is common in many metals, where no specific chemical bonds require any other, less condensed arrangements. Since each atom is spherically symmetrical, close packing of atoms is equivalent to arranging spheres in the minimal space. This can be achieved in various ways.

First, in a single plane the most condensed arrangement of circles possible is that of Fig. 6.25*a*, in which each circle touches six other circles. To extend this to three dimensions we can place another equivalent layer above the first, as in Fig. 6.25*b*. There are two ways of placing the second layer over the first one, but they are completely equivalent and physically indistinguishable. They can be seen if we place an asterisk in alternate interstices in the first layer so that no two starred nor two unstarred interstices have a common corner; then the second layer, which will pack closest onto the first, must have spheres either above each of the starred, or, as in Fig. 6.25*b*, each unstarred interstice of the original layer. Each sphere now

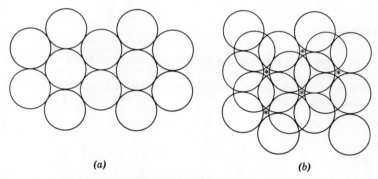

(*a*)   (*b*)

Fig. 6.25   Close packing: (*a*) one layer, (*b*) two layers.

touches six other spheres on the same plane and three in the other, a total of nine. We can now place a third layer of spheres over the two. This can be done in two ways, which now are distinct. The new layer can be placed so that the spheres lie directly above the *starred interstices* of the first layer, or so that they lie directly above the *spheres* of the first layer; in either case each sphere of the central layer touches twelve others.

Each close packed layer has the symmetry *6mm*. The two-layer array of Fig. 6.25*b* has lower symmetry, *3m*; this is the lowest symmetry that an array of any number of layers can have. However, higher symmetry is possible if a symmetric repetition of layer types occurs. The sequence in which the third layer lies above the interstices of the first has the symmetry *m3m* (*4/m 3 2/m, O_h*), and an infinite repetition gives *cubic close packing*, belonging to space group *F m3m*. All other possible sequences of layers have hexagonal symmetry; of these, alternation of layers is the most common, that is, an infinite repetition of Fig. 6.25*b*. It is called hexagonal close packing and belongs to space group $P\,6_3/m\,mc$. Both types are encountered in the structure of metals.

Other packing arrangements of spheres are, of course, possible; spheres of equal size may pack in a more open arrangement with fewer neighbors, say eight, six, or even four. When the spheres have different sizes, like atoms of different kinds, each with its own radius, many other packings must occur. We will not pursue this further.

REFERENCES

P. J. Wheatley, *The Determination of Molecular Structure*, Oxford University Press, London, 1959.

L. V. Azároff, *Introduction to Solids*, McGraw-Hill Book Co., New York, 1960, Chapters 1 and 2.

Appendix 1

Character Tables

This appendix lists the character tables for all point groups commonly encountered in real molecules. The construction of these character tables was considered in Chapter 4. The first column of each table lists the various symmetry species applicable to the particular point group; the body lists the characters for each of the important symmetry operations in columns headed by the operations, and the next to final column lists the three coordinate axes (x,y,z), which can stand for the translation vectors and for the components of the dipole moment vector, and the three rotations, R_x, R_y, and R_z, in the rows of the symmetry species to which they belong. The final column lists the six components of the polarizability tensor. The arrangement of the character tables follows Herzberg.[1]

Table A1.1 Symmetry Species and Characters for the Point Groups C_2, C_s, $C_i = S_2$

C_2	I	$C_2(z)$		
A	$+1$	$+1$	z, R_z	$\alpha_{xx}, \alpha_{yy}, \alpha_{zz}, \alpha_{xy}$
B	$+1$	-1	x, y, R_x, R_y	α_{xz}, α_{yz}

C_s	I	$\sigma(xy)$		
A'	$+1$	$+1$	x, y, R_z	$\alpha_{xx}, \alpha_{yy}, \alpha_{zz}, \alpha_{xy}$
A''	$+1$	-1	z, R_x, R_y	α_{xz}, α_{yz}

$C_i \equiv S_2$	I	i	
A_g	$+1$	$+1$	R_x, R_y, R_z; all α
A_u	$+1$	-1	x, y, z

[1] G. Herzberg, *Molecular Spectra and Molecular Structure*, D. Van Nostrand Co., Princeton, N.J., 1945.

159

Table A1.2 Symmetry Species and Characters for the Point Groups C_{2v}, C_{2h}, and $D_2 \equiv V$

C_{2v}	I	$C_2(z)$	$\sigma_v(xz)$	$\sigma_v(yz)$		
A_1	$+1$	$+1$	$+1$	$+1$	z	$\alpha_{xx}, \alpha_{yy}, \alpha_{zz}$
A_2	$+1$	$+1$	-1	-1	R_z	α_{xy}
B_1	$+1$	-1	$+1$	-1	x, R_y	α_{xz}
B_2	$+1$	-1	-1	$+1$	y, R_x	α_{yz}

C_{2h}	I	$C_2(z)$	$\sigma_h(xy)$	i		
A_g	$+1$	$+1$	$+1$	$+1$	R_z	$\alpha_{xx}, \alpha_{yy}, \alpha_{zz}, \alpha_{xy}$
A_u	$+1$	$+1$	-1	-1	z	—
B_g	$+1$	-1	-1	$+1$	R_x, R_y	α_{xz}, α_{yz}
B_u	$+1$	-1	$+1$	-1	x, y	—

$D_2 \equiv V$	I	$C_2(z)$	$C_2(y)$	$C_2(x)$		
A	$+1$	$+1$	$+1$	$+1$	—	$\alpha_{xx}, \alpha_{yy}, \alpha_{zz}$
B_1	$+1$	$+1$	-1	-1	z, R_z	α_{xy}
B_2	$+1$	-1	$+1$	-1	y, R_y	α_{xz}
B_3	$+1$	-1	-1	$+1$	x, R_x	α_{yz}

Table A1.3 Symmetry Species and Characters for the Point Group $D_{2h} \equiv V_h$

$D_{2h} \equiv V_h$	I	$\sigma(xy)$	$\sigma(xz)$	$\sigma(yz)$	i	$C_2(z)$	$C_2(y)$	$C_2(x)$		
A_g	$+1$	$+1$	$+1$	$+1$	$+1$	$+1$	$+1$	$+1$	—	$\alpha_{xx}, \alpha_{yy}, \alpha_{zz}$
A_u	$+1$	-1	-1	-1	-1	$+1$	$+1$	$+1$	—	—
B_{1g}	$+1$	$+1$	-1	-1	$+1$	$+1$	-1	-1	R_z	—
B_{1u}	$+1$	-1	$+1$	$+1$	-1	$+1$	-1	-1	z	α_{xy}
B_{2g}	$+1$	-1	$+1$	-1	$+1$	-1	$+1$	-1	R_y	—
B_{2u}	$+1$	$+1$	-1	$+1$	-1	-1	$+1$	-1	y	α_{xz}
B_{3g}	$+1$	-1	-1	$+1$	$+1$	-1	-1	$+1$	R_x	—
B_{3u}	$+1$	$+1$	$+1$	-1	-1	-1	-1	$+1$	x	α_{yz}

Table A1.4 Symmetry Species and Characters for the Point Group C_3

C_3	I	$2C_3$		
A	$+1$	$+1$	z, R_z	$\alpha_{xx} + \alpha_{yy}, \alpha_{zz}$
E	$+2$	-1	x, y, R_x, R_y	$\alpha_{xy}, \alpha_{yz}, \alpha_{yz}, \alpha_{xx} - \alpha_{yy}$

Table A1.5 Symmetry Species and Characters for the Point Group C_{3h}

C_{3h}	I	$2C_3$	σ_h	$2S_3$		
A'	$+1$	$+1$	$+1$	$+1$	R_z	$\alpha_{xx}+\alpha_{yy},\ \alpha_{zz}$
A''	$+1$	$+1$	-1	-1	z	—
E'	$+2$	-1	$+2$	-1	x,y	$\alpha_{xx}-\alpha_{yy},\ \alpha_{xy}$
E''	$+2$	-1	-2	$+1$	R_x, R_y	$\alpha_{xz},\ \alpha_{yz}$

Table A1.6 Symmetry Species and Characters for the Point Groups C_{3v} and D_3

C_{3v}	I	$2C_3(z)$	$3\sigma_v$		
A_1	$+1$	$+1$	$+1$	z	$\alpha_{xx}+\alpha_{yy},\ \alpha_{zz}$
A_2	$+1$	$+1$	-1	R_z	$\alpha_{xx}-\alpha_{yy},\ \alpha_{xy},\ \alpha_{xz},\ \alpha_{yz}$
E	$+2$	-1	0	x, y, R_x, R_y	—

D_3	I	$2C_3(z)$	$3C_2$		
A_1	$+1$	$+1$	$+1$	—	$\alpha_{xx}+\alpha_{yy},\ \alpha_{zz}$
A_2	$+1$	$+1$	-1	z, R_z	—
E	$+2$	-1	0	x, y, R_x, R_y	$\alpha_{xx}-\alpha_{yy},\ \alpha_{xy},\ \alpha_{yz},\ \alpha_{xz}$

Table A1.7 Symmetry Species and Characters for the Point Group D_{3h}

D_{3h}	I	$2C_3(z)$	$3C_2$	σ_h	$2S_3$	$3\sigma_v$		
A_1'	$+1$	$+1$	$+1$	$+1$	$+1$	$+1$	—	$\alpha_{xx}+\alpha_{yy},\ \alpha_{zz}$
A_1''	$+1$	$+1$	$+1$	-1	-1	-1	—	—
A_2'	$+1$	$+1$	-1	$+1$	$+1$	-1	R_z	—
A_2''	$+1$	$+1$	-1	-1	-1	$+1$	z	—
E'	$+2$	-1	0	$+2$	-1	0	x, y	$\alpha_{xx}-\alpha_{yy},\ \alpha_{xy}$
E''	$+2$	-1	0	-2	$+1$	0	R_x, R_y	$\alpha_{xz},\ \alpha_{yz}$

Table A1.8 Symmetry Species and Characters for the Point Group $D_{3d}(\equiv S_{6v})$

D_{3d}	I	$2S_6(z)$	$2S_6{}^2\equiv 2C_3$	$S_6{}^3\equiv S_2\equiv i$	$3C_2$	$3\sigma_d$		
A_{1g}	$+1$	$+1$	$+1$	$+1$	$+1$	$+1$	—	$\alpha_{xx}+\alpha_{yy},\ \alpha_{zz}$
A_{1u}	$+1$	-1	$+1$	-1	$+1$	-1	—	—
A_{2g}	$+1$	$+1$	$+1$	$+1$	-1	-1	R_z	—
A_{2u}	$+1$	-1	$+1$	-1	-1	$+1$	z	—
E_g	$+2$	-1	-1	$+2$	0	0	R_x, R_y	$\alpha_{xx}-\alpha_{yy},\ \alpha_{xy},\ \alpha_{xz},\ \alpha_{yz}$
E_u	$+2$	$+1$	-1	-2	0	0	x, y	—

Table A1.9   Symmetry Species and Characters of the Point Groups C_4 and S_4

C_4	I	$2C_4$	$C_4{}^2 \equiv C_2$		
S_4	I	$2S_4$	$S_4{}^2 \equiv C_2$		
A	$+1$	$+1$	$+1$	z for C_4, Rz	$\alpha_{xx} + \alpha_{yy}, \alpha_{zz}$
B	$+1$	-1	$+1$	z for S_4	$\alpha_{xx} - \alpha_{yy}, \alpha_{xy}$
E	$+2$	0	-2	x, y, R_x, R_y	α_{xz}, α_{yz}

Table A1.10   Symmetry Species and Characters for the Point Groups C_{4v}, D_4, and $D_{2d} \equiv V_d$

C_{4v}	I	$2C_4(z)$	$C_4{}^2 \equiv C_2''$	$2\sigma_v$	$2\sigma_d$		
D_4	I	$2C_4(z)$	$C_4{}^2 \equiv C_2''$	$2C_2$	$2C_2'$		
$D_{2d} \equiv V_d \equiv S_4$	I	$2S_4(z)$	$S_4{}^2 \equiv C_2''$	$2C_2$	$2\sigma_d$		
A_1	$+1$	$+1$	$+1$	$+1$	$+1$	z for C_{4v}	$\alpha_{xx} + \alpha_{yy}, \alpha_{zz}$
A_2	$+1$	$+1$	$+1$	-1	-1	z for D_4, R_z	—
B_1	$+1$	-1	$+1$	$+1$	-1	—	$\alpha_{xx} - \alpha_{yy}$
B_2	$+1$	-1	$+1$	-1	$+1$	z for V_d	α_{xy}
E	$+2$	0	-2	0	0	x, y, R_x, R_y	α_{xz}, α_{yz}

Table A1.11   Symmetry Species and Characters for the Point Group C_{4h}

C_{4h}	I	$2C_4$	$C_4{}^2 \equiv C_2''$	σ_h	$2S_4$	$S_2 \equiv i$		
A_g	$+1$	$+1$	$+1$	$+1$	$+1$	$+1$	R_z	$\alpha_{xx} + \alpha_{yy}, \alpha_{zz}$
A_u	$+1$	$+1$	$+1$	-1	-1	-1	z	—
B_g	$+1$	-1	$+1$	$+1$	-1	$+1$	—	$\alpha_{xx} - \alpha_{yy}, \alpha_{xy}$
B_u	$+1$	-1	$+1$	-1	$+1$	-1	—	—
E_g	$+2$	0	-2	-2	0	$+2$	R_x, R_y	α_{xz}, α_{yz}
E_u	$+2$	0	-2	$+2$	0	-2	x, y	—

Table A1.12   Symmetry Species and Characters for the Point Group $D_{4d}(\equiv S_{8v})$

D_{4d}	I	$2S_8(z)$	$2S_8{}^2 \equiv 2C_4$	$2S_8{}^3$	$S_8{}^4 \equiv C_2''$	$4C_2$	$4\sigma_d$		
A_1	$+1$	$+1$	$+1$	$+1$	$+1$	$+1$	$+1$	—	$\alpha_{xx} + \alpha_{yy}, \alpha_{zz}$
A_2	$+1$	$+1$	$+1$	$+1$	$+1$	-1	-1	R_z	—
B_1	$+1$	-1	$+1$	-1	$+1$	$+1$	-1	—	—
B_2	$+1$	-1	$+1$	-1	$+1$	-1	$+1$	z	—
E_1	$+2$	$+\sqrt2$	0	$-\sqrt2$	-2	0	0	x, y	—
E_2	$+2$	0	-2	-0	$+2$	0	0	—	$\alpha_{xx} - \alpha_{yy}, \alpha_{xy}$
E_3	$+2$	$-\sqrt2$	0	$+\sqrt2$	-2	0	0	R_x, R_y	α_{xz}, α_{yz}

Table A1.13   Symmetry Species and Characters for the Point Group D_{4h}

D_{4h}	I	$2C_4(z)$	$C_4{}^2 \equiv C_2''$	$2C_2$	$2C_2'$	σ_h	$2\sigma_v$	$2\sigma_d$	$2S_4$	$S_2 \equiv i$		
A_{1g}	+1	+1	+1	+1	+1	+1	+1	+1	+1	+1	—	$\alpha_{xx}+\alpha_{yy}, \alpha_{zz}$
A_{1u}	+1	+1	+1	+1	+1	−1	−1	−1	−1	−1	—	—
A_{2g}	+1	+1	+1	−1	−1	+1	−1	−1	+1	+1	R_z	—
A_{2u}	+1	+1	+1	−1	−1	−1	+1	+1	−1	−1	z	$\alpha_{xx}-\alpha_{yy}$
B_{1g}	+1	−1	+1	+1	−1	+1	+1	−1	−1	+1	—	
B_{1u}	+1	−1	+1	+1	−1	−1	−1	+1	+1	−1	—	—
B_{2g}	+1	−1	+1	−1	+1	+1	−1	+1	−1	+1	—	α_{xy}
B_{2u}	+1	−1	+1	−1	+1	−1	+1	−1	+1	−1	—	—
E_g	+2	0	−2	0	0	−2	0	0	0	+2	R_x, R_y	α_{xz}, α_{yz}
E_u	+2	0	−2	0	0	+2	0	0	0	−2	x, y	

Table A1.14   Symmetry Species and Characters for the Point Groups C_5, C_{5v}, $D_5{}^a$

C_{5v}	I	$2C_5$	$2C_5{}^2$	$5\sigma_v{}^b$		
A_1	+1	+1	+1	+1	z for C_{5v}	$\alpha_{xx}+\alpha_{yy}, \alpha_{zz}$
A_2	+1	+1	+1	−1	z for D_5, R_z	—
E_1	+2	$2\cos 72°$	$2\cos 144°$	0	x, y, R_x, R_y	α_{xz}, α_{yz}
E_2	+2	$2\cos 144°$	$2\cos 72°$	0	—	$\alpha_{xx}-\alpha_{yy}, \alpha_{xy}$

^{a} In C_5, A_1 and A_2 coalesce to form A, since there is no σ_v.
^{b} In D_5, replace $5\sigma_v$ by $5C_2$.

Table A1.15   Symmetry Species and Characters for the Point Group C_{5h}

C_{5h}	I	$2C_5$	$2C_5{}^2$	σ_h	$2S_5{}^3$	$2S_5{}^2$		
A'	+1	+1	+1	+1	1	1	R_z	$\alpha_{xx}+\alpha_{yy}, \alpha_{zz}$
A''	+1	+1	+1	−1	−1	−1	z	—
E_1'	+2	$2\cos 72°$	$2\cos 144°$	+2	$2\cos 72°$	$2\cos 144°$	x, y	—
E_1''	+2	$2\cos 72°$	$2\cos 144°$	−2	$-2\cos 72°$	$-2\cos 144°$	R_x, R_y	α_{xz}, α_{yz}
E_2'	+2	$2\cos 144°$	$2\cos 72°$	+2	$2\cos 144°$	$2\cos 72°$	—	$\alpha_{xx}-\alpha_{yy}, \alpha_{xy}$
E_2''	+2	$2\cos 144°$	$2\cos 72°$	−2	$-2\cos 144°$	$-2\cos 72°$	—	—

Table A1.16 Symmetry Species and Characters for the Point Groups D_{5h} and D_{5d}

D_{5d}	—	I	$2C_5$	$2C_5^2$	i	$5C_2$	$5\sigma_d$	$2S_{10}^3$	$2S_{10}$		
	D_{5h}	I	$2C_5$	$2C_5^2$	σ_h	$5C_2$	$5\sigma_v$	$2S_5$	$2S_5^3$		
A_{1g}	A_1'	$+1$	$+1$	$+1$	$+1$	$+1$	$+1$	$+1$	$+1$		$\alpha_{xx} + \alpha_{yy},\ \alpha_{zz}$
A_{1u}	A_1''	$+1$	$+1$	$+1$	-1	$+1$	-1	-1	-1		—
A_{2g}	A_2'	$+1$	$+1$	$+1$	$+1$	-1	-1	$+1$	$+1$	R_z	—
A_{2u}	A_2''	$+1$	$+1$	$+1$	-1	-1	$+1$	-1	-1	z	—
E_{1g}	E_1'	$+2$	$2\cos 72°$	$2\cos 144°$	$+2$	0	0	$+2\cos 72°$	$+2\cos 144°$	R_x, R_y	$(\alpha_{xz}, \alpha_{yz})$ for D_{5d}
E_{1u}	E_1''	$+2$	$2\cos 72°$	$2\cos 144°$	-2	0	0	$-2\cos 72°$	$-2\cos 144°$	x, y	$(\alpha_{xz}, \alpha_{yz})$ for D_{5h}
E_{2g}	E_2'	$+2$	$2\cos 144°$	$2\cos 72°$	$+2$	0	0	$+2\cos 144°$	$+2\cos 72°$	—	$\alpha_{xx} - \alpha_{yy}$
E_{2u}	E_2''	$+2$	$2\cos 144°$	$2\cos 72°$	-2	0	0	$-2\cos 144°$	$-2\cos 72°$	—	—

Table A1.17 Symmetry Species and Characters for the Point Groups C_6 and C_{6h}[a]

C_{6h}	I	$2C_6$	$2C_6{}^2 \equiv C_3$	$C_6{}^3 \equiv C_2''$	σ_h	$2S_6$	$2S_3$	$S_2 \equiv i$		
A_g	$+1$	$+1$	$+1$	$+1$	$+1$	$+1$	$+1$	$+1$	R_z	$\alpha_{xx}+\alpha_{yy}, \alpha_{zz}$
A_u	$+1$	$+1$	$+1$	$+1$	-1	-1	-1	-1	z	—
B_g	$+1$	-1	$+1$	-1	-1	$+1$	-1	$+1$	—	—
B_u	$+1$	-1	$+1$	-1	$+1$	-1	$+1$	-1	—	—
E_{1g}	$+2$	$+1$	-1	-2	-2	-1	$+1$	$+2$	R_x, R_y	α_{xz}, α_{yz}
E_{1u}	$+2$	$+1$	-1	-2	$+2$	$+1$	-1	-2	x, y	—
E_{2g}	$+2$	-1	-1	$+2$	$+2$	-1	-1	$+2$	—	$\alpha_{xx}-\alpha_{yy}, \alpha_{xy}$
E_{2u}	$+2$	-1	-1	$+2$	-2	$+1$	$+1$	-2	—	

[a] For C_6, absence of σ_h, S_6, S_3 and $S_2 \equiv i$ eliminates the g, u classification and reduces the species to A, B, E_1, and E_2.

Table A1.18 Symmetry Species and Characters for the Point Groups C_{6v} and D_6

C_{6v}	I	$2C_6(z)$	$2C_6{}^2 \equiv 2C_3$	$C_6{}^3 \equiv C_2''$	$3\sigma_v$	$3\sigma_d$		
D_6	I	$2C_6(z)$	$2C_6{}^2 \equiv 2C_3$	$C_6{}^3 \equiv C_2''$	$3C_2$	$3C_2'$		
A_1	$+1$	$+1$	$+1$	$+1$	$+1$	$+1$	z for C_{6v}	$\alpha_{xx}+\alpha_{yy}, \alpha_{zz}$
A_2	$+1$	$+1$	$+1$	$+1$	-1	-1	z for D_6, R_z	—
B_1	$+1$	-1	$+1$	-1	$+1$	-1	—	—
B_2	$+1$	-1	$+1$	-1	-1	$+1$	—	—
E_1	$+2$	$+1$	-1	-2	0	0	x, y, R_x, R_y	α_{xz}, α_{yz}
E_2	$+2$	-1	-1	$+2$	0	0	—	$\alpha_{xx}-\alpha_{yy}, \alpha_{xy}$

Table A1.19 Symmetry Species and Characters for the Point Group D_{6h}

D_{6h}	I	$2C_6(z)$	$2C_6^2 \equiv 2C_3$	$C_6^3 \equiv C_2''$	$3C_2$	$3C_2'$	σ_h	$3\sigma_v$	$3\sigma_d$	$2S_6$	$2S_3$	$S_6^3 \equiv S_2 \equiv i$		
A_{1g}	+1	+1	+1	+1	+1	+1	+1	+1	+1	+1	+1	+1	—	$\alpha_{xx}+\alpha_{yy},\ \alpha_{zz}$
A_{1u}	+1	+1	+1	+1	+1	+1	−1	−1	−1	−1	−1	−1	—	—
A_{2g}	+1	+1	+1	+1	−1	−1	+1	−1	−1	+1	+1	+1	R_z	—
A_{2u}	+1	+1	+1	+1	−1	−1	−1	+1	+1	−1	−1	−1	z	—
B_{1g}	+1	−1	+1	−1	+1	−1	−1	−1	+1	+1	−1	+1	—	—
B_{1u}	+1	−1	+1	−1	+1	−1	+1	+1	−1	−1	+1	−1	—	—
B_{2g}	+1	−1	+1	−1	−1	+1	−1	+1	−1	+1	−1	+1	—	—
B_{2u}	+1	−1	+1	−1	−1	+1	+1	−1	+1	−1	+1	−1	—	—
E_{1g}	+2	+1	−1	−2	0	0	−2	0	0	−1	+1	+2	R_x, R_y	α_{xz}, α_{yz}
E_{1u}	+2	+1	−1	−2	0	0	+2	0	0	+1	−1	−2	x, y	—
E_{2g}	+2	−1	−1	+2	0	0	+2	0	0	−1	−1	+2	—	$\alpha_{xx}-\alpha_{yy},\ \alpha_{xy}$
E_{2u}	+2	−1	−1	+2	0	0	−2	0	0	+1	+1	−2	—	—

Table A1.20 Symmetry Species and Characters for the Point Group T

T	I	$8C_3$	$3C_2$		
A	+1	+1	+1	—	$\alpha_{xx}+\alpha_{yy}+\alpha_{zz}$
E	+2	−1	+2	—	$\alpha_{xx}+\alpha_{yy}-2\alpha_{zz},\ \alpha_{xx}-\alpha_{yy}$
T	+3	0	−1	x, y, z, R_x, R_y, R_z	$\alpha_{xy}, \alpha_{xz}, \alpha_{yz}$

Table A1.21 Symmetry Species and Characters for the Point Groups T_d and O

T_d	I	$8C_3$	$6\sigma_d$	$6S_4$	$3S_4^2 \equiv 3C_2$		
O	I	$8C_3$	$6C_3$	$6C_2$	$3C_4^2 \equiv 3C_2''$		
A_1	+1	+1	+1	+1	+1		$\alpha_{xx}+\alpha_{yy}+\alpha_{zz}$
A_2	+1	+1	-1	-1	+1		—
E	+2	-1	0	0	+2		$\alpha_{zz}+\alpha_{yy}-2\alpha_{zz},\ \alpha_{zz}-\alpha_{yy}$
T_1	+3	0	-1	+1	-1	x,y,z for O, R_x, R_y, R_z	—
T_2	+3	0	+1	-1	-1	x,y,z for T_d	$\alpha_{xy},\alpha_{zz},\alpha_{yz}$

Table A1.22 Symmetry Species and Characters for the Point Group O_h

O_h	I	$8C_3$	$6C_2$	$6C_4$	$3C_4^2 \equiv 3C_2$	$S_2 \equiv i$	$6S_4$	$8S_6$	$3\sigma_h$	$6\sigma_d$		
A_{1g}	+1	+1	+1	+1	+1	+1	+1	+1	+1	+1		$\alpha_{xx}+\alpha_{yy}+\alpha_{zz}$
A_{1u}	+1	+1	+1	+1	+1	-1	-1	-1	-1	-1		—
A_{2g}	+1	+1	-1	-1	+1	+1	-1	+1	+1	-1		—
A_{2u}	+1	+1	-1	-1	+1	-1	+1	-1	-1	+1		—
E_g	+2	-1	0	0	+2	+2	0	-1	+2	0		$\alpha_{xx}+\alpha_{yy}+2\alpha_{zz},\ \alpha_{xx}-\alpha_{yy}$
E_u	+2	-1	0	0	+2	-2	0	+1	-2	0		—
T_{1g}	+3	0	-1	+1	-1	+3	+1	0	-1	-1	R_x, R_y, R_z	—
T_{1u}	+3	0	-1	+1	-1	-3	-1	0	+1	+1	x, y, z	—
T_{2g}	+3	0	+1	-1	-1	+3	-1	0	-1	+1		$\alpha_{xy},\alpha_{xz},\alpha_{yz}$
T_{2u}	+3	0	+1	-1	-1	-3	+1	0	+1	-1		—

Table A1.23   Symmetry Species and Character Tables for the Point Groups I and $I_h{}^a$

I_h	E	$12C_5$	$12C_5^2$	$20C_3$	$15C_2$	i	$12S_{10}$	$12S_{10}^3$	$20S_6$	15σ	
A_g	1	1	1	1	1	1	1	1	1	1	
A_u	1	1	1	1	1	-1	-1	-1	-1	-1	
T_{1g}	3	$(1+\sqrt5)/2$	$(1-\sqrt5)/2$	0	-1	3	$(1-\sqrt5)/2$	$(1+\sqrt5)/2$	0	-1	R_x, R_y, R_z
T_{1u}	3	$(1+\sqrt5)/2$	$(1-\sqrt5)/2$	0	-1	-3	$-(1-\sqrt5)/2$	$-(1+\sqrt5)/2$	0	1	x, y, z
T_{2g}	3	$(1-\sqrt5)/2$	$(1+\sqrt5)/2$	0	-1	3	$(1+\sqrt5)/2$	$(1-\sqrt5)/2$	0	-1	
T_{2u}	3	$(1-\sqrt5)/2$	$(1+\sqrt5)/2$	0	-1	-3	$-(1+\sqrt5)/2$	$-(1-\sqrt5)/2$	0	1	
$G_g{}^b$	4	-1	-1	1	0	4	-1	-1	1	0	
$G_u{}^b$	4	-1	-1	1	0	-4	1	1	-1	0	
$H_g{}^c$	5	0	0	-1	1	5	0	0	-1	1	
$H_u{}^c$	5	0	0	-1	1	-5	0	0	1	-1	

^{a} For I, absence of i, the S_{10}, S_6, and σ eliminates the g, u classification, and leaves species A, T_1, T_2, G, and H.

^{b} A fourfold degenerate species.

^{c} A fivefold degenerate species.

Table A1.24 Symmetry Species and Characters for the Point Group $C_{\infty v}$

$C_{\infty v}$	I	$2C_\infty^{\varphi}$	$2C_\infty^{2\varphi}$	$2C_\infty^{3\varphi}$		$\infty\sigma_v$		
Σ^+	$+1$	$+1$	$+1$	$+1$		$+1$	z	$\alpha_{xx}+\alpha_{yy},\ \alpha_{zz}$
Σ^-	$+1$	$+1$	$+1$	$+1$		-1	R_z	—
Π	$+2$	$2\cos\varphi$	$2\cos 2\varphi$	$2\cos 3\varphi$		0	x,y,R_x,R_y	α_{xz},α_{yz}
Δ	$+2$	$2\cos 2\varphi$	$2\cos 2\cdot2\varphi$	$2\cos 2\cdot3\varphi$		0	—	$\alpha_{xx}-\alpha_{yy},\alpha_{xy}$
Φ	$+2$	$2\cos 3\varphi$	$2\cos 2\cdot3\varphi$	$2\cos 3\cdot3\varphi$		0	—	—
	—							

Table A1.25 Symmetry Species and Characters for the Point Group $D_{\infty h}$

$D_{\infty h}$	I	$2C_\infty^{\varphi}$	$2C_\infty^{2\varphi}$	$2C_\infty^{3\varphi}$		σ_h	∞C_2	$\infty\sigma_v$	$2S_\infty^{\varphi}$	$2S_\infty^{2\varphi}$		$S_2\equiv i$		
Σ_g^+	$+1$	$+1$	$+1$	$+1$		$+1$	$+1$	$+1$	$+1$	$+1$		$+1$	—	$\alpha_{xx}+\alpha_{yy},\ \alpha_{zz}$
Σ_u^+	$+1$	$+1$	$+1$	$+1$		-1	-1	$+1$	-1	-1		-1	z	—
Σ_g^-	$+1$	$+1$	$+1$	$+1$		$+1$	-1	-1	$+1$	$+1$		$+1$	R_z	—
Σ_u^-	$+1$	$+1$	$+1$	$+1$		-1	$+1$	-1	-1	-1		-1	—	—
Π_g	$+2$	$2\cos\varphi$	$2\cos 2\varphi$	$2\cos 3\varphi$		-2	0	0	$-2\cos\varphi$	$-2\cos 2\varphi$		$+2$	R_x,R_y	α_{xz},α_{yz}
Π_u	$+2$	$2\cos\varphi$	$2\cos 2\varphi$	$2\cos 3\varphi$		$+2$	0	0	$+2\cos\varphi$	$+2\cos 2\varphi$		-2	x,y	—
Δ_g	$+2$	$2\cos 2\varphi$	$2\cos 4\varphi$	$2\cos 6\varphi$		$+2$	0	0	$+2\cos 2\varphi$	$+2\cos 4\varphi$		$+2$	—	$\alpha_{xx}-\alpha_{yy},\alpha_{xy}$
Δ_u	$+2$	$2\cos 2\varphi$	$2\cos 4\varphi$	$2\cos 6\varphi$		-2	0	0	$-2\cos 2\varphi$	$-2\cos 4\varphi$		-2	—	—
Φ_g	$+2$	$2\cos 3\varphi$	$2\cos 6\varphi$	$2\cos 9\varphi$		-2	0	0	$-2\cos 3\varphi$	$-2\cos 4\varphi$		$+2$	—	—
Φ_u	$+2$	$2\cos 3\varphi$	$2\cos 6\varphi$	$2\cos 9\varphi$		$+2$	0	0	$+2\cos 3\varphi$	$+2\cos 4\varphi$		-2	—	—
	—													

Appendix 2

The Number of Normal Vibrations in Various Symmetry Species

Point Group, Total Number of Atoms	Species of Vibration	Number of Vibrations[a]
C_2 ($N = 2m + m_0$)	A B	$3m + m_0 - 2$ $3m + 2m_0 - 4$
$C_s \equiv C_{1h}$ ($N = 2m + m_0$)	A' A''	$3m + 2m_0 - 3$ $3m + m_0 - 3$
$C_i \equiv S_2$ ($N = 2m + m_0$)	A_g A_u	$3m - 3$ $3m + 3m_0 - 3$
C_{2v} ($N = 4m + 2m_{xz} + 2m_{yz} + m_0$)	A_1 A_2 B_1 B_2	$3m + 2m_{xz} + 2m_{yz} + m_0 - 1$ $3m + m_{xz} + m_{yz} - 1$ $3m + 2m_{xz} + m_{yz} + m_0 - 2$ $3m + m_{xz} + 2m_{yz} + m_0 - 2$
C_{2h} ($N = 4m + 2m_h + 2m_2 + m_0$)	A_g A_u B_g B_u	$3m + 2m_h + m_2 - 1$ $3m + m_h + m_2 + m_0 - 1$ $3m + m_h + 2m_2 - 2$ $3m + 2m_h + 2m_2 + 2m_0 - 2$

$$D_2 \equiv V$$
$$(N = 4m + 2m_{2x} + 2m_{2y} + 2m_{2z} + m_0)$$

A	$3m + m_{2x} + m_{2y} + m_{2z}$
B_1	$3m + 2m_{2x} + 2m_{2y} + m_{2z} + m_0 - 2$
B_2	$3m + 2m_{2x} + m_{2y} + 2m_{2z} + m_0 - 2$
B_3	$3m + m_{2x} + 2m_{2y} + 2m_{2z} + m_0 - 2$

$$D_{2h} \equiv V_h$$
$$(N = 8m + 4m_{xy} + 4m_{xz} + 4m_{yz} + 2m_{2x} + 2m_{2y} + 2m_{2z} + m_0)$$

A_g	$3m + 2m_{xy} + 2m_{xz} + 2m_{yz} + m_{2x} + m_{2y} + m_{2z}$
A_u	$3m + m_{xy} + m_{xz} + m_{yz}$
B_{1g}	$3m + 2m_{xy} + m_{xz} + m_{yz} + m_{2x} + m_{2y} - 1$
B_{1u}	$3m + m_{xy} + 2m_{xz} + 2m_{yz} + m_{2x} + m_{2y} + m_0 - 1$
B_{2g}	$3m + m_{xy} + 2m_{xz} + m_{yz} + m_{2x} + m_{2z} - 1$
B_{2u}	$3m + 2m_{xy} + m_{xz} + 2m_{yz} + m_{2x} + m_{2z} + m_0 - 1$
B_{3g}	$3m + m_{xy} + m_{xz} + 2m_{yz} + m_{2y} + m_{2z} - 1$
B_{3u}	$3m + 2m_{xy} + 2m_{xz} + m_{yz} + m_{2y} + m_{2z} + m_0 - 1$

[a] m is the number of sets of nuclei not on any element of symmetry; m_0 is the number of nuclei on all elements of symmetry; $m_2, m_3, m_4, \ldots$ are the numbers of sets of nuclei on a twofold, threefold, fourfold, ... axis but not on any other element of symmetry that does not wholly coincide with that axis; m_2' is the number of sets of nuclei on a twofold axis called C_2' in the character tables; m_v, m_d, m_h are the numbers of sets of nuclei on planes $\sigma_v, \sigma_d, \sigma_h$, respectively, but not on any other element of symmetry. If necessary, various m_2 are distinguished as m_{2x}, m_{2y}, and m_{2z}, referring to C_2^x, C_2^y, and C_2^z, respectively; similarly, m_{xy}, m_{xz}, and m_{yz} refer to σ^{xy}, σ^{xz}, and σ^{yz}, respectively.

Point Group, Total Number of Atoms	Species of Vibration	Number of Vibrations
C_3 $(N = 3m + m_0)$	A	$3m + m_0 - 2$
	E	$3m + m_0 - 2$
C_4 $(N = 4m + m_0)$	A	$3m + m_0 - 2$
	B	$3m$
	E	$3m + m_0 - 2$
C_6 $(N = 6m + m_0)$	A	$3m + m_0 - 2$
	B	$3m$
	E_1	$3m + m_0 - 2$
	E_2	$3m$
S_4 $(N = 4m + 2m_2 + m_0)$	A	$3m + m_2 - 1$
	B	$3m + m_2 + m_0 - 1$
	E	$3m + 2m_2 + m_0 - 2$
S_6 $(B = 6m + 2m_3 + m_0)$	A_g	$3m + m_2 - 1$
	B_u	$3m + m_3 + m_0 - 1$
	E_{1u}	$3m + m_3 + m_0 - 1$
	E_{2g}	$3m + m_3 - 1$
D_3 $(N = 6m + 3m_2 + 2m_3 + m_0)$	A_1	$3m + m_2 + m_3$
	A_2	$3m + 2m_2 + m_3 + m_0 - 2$
	E	$6m + 3m_2 + 2m_3 + m_0 - 2$

Point Group, Total Number of Atoms	Species of Vibration	Number of Vibrations
D_4 $(N = 8m + 4m_2 + 4m_2'$ $+ 2m_4 + m_0)$	A_1	$3m + m_2 + m_2' + m_4$
	A_2	$3m + 2m_2 + 2m_2' + m_4 + m_0 - 2$
	B_1	$3m + m_2 + 2m_2'$
	B_2	$3m + 2m_2 + m_2'$
	E	$6m + 3m_2 + 3m_2' + 2m_4 + m_0 - 2$
D_6 $(N = 12m + 6m_2 + 6m_2'$ $+ 2m_6 + m_0)$	A_1	$3m + m_2 + m_2' + m_6$
	A_2	$3m + 2m_2 + 2m_2' + m_6 + m_0 - 2$
	B_1	$3m + m_2 + 2m_2'$
	B_2	$3m + 2m_2 + m_2'$
	E_1	$6m + 3m_2 + 3m_2' + 2m_6 + m_0 - 2$
	E_2	$6m + 3m_2 + 3m_2'$
C_{3v} $(N = 6m + 3m_v + m_0)$	A_1	$3m + 2m_v + m_0 - 1$
	A_2	$3m + m_v - 1$
	E	$6m + 3m_v + m_0 - 2$
C_{4v} $(N = 8m + 4m_v + 4m_d + m_0)$	A_1	$3m + 2m_v + 2m_d + m_0 - 1$
	A_2	$3m + m_v + m_d - 1$
	B_1	$3m + 2m_v + m_d$
	B_2	$3m + m_v + 2m_d$
	E	$6m + 3m_v + 3m_d + m_0 - 2$

Point Group, Total Number of Atoms	Species of Vibration	Number of Vibrations
C_{5v} ($N = 10m + 5m_v + m_0$)	A_1	$3m + 2m_v + m_0 - 1$
	A_2	$3m + m_v - 1$
	E_1	$6m + 3m_v + m_0 - 2$
	E_2	$6m + 3m_v$
C_{6v} ($N = 12m + 6m_v + 6m_d + m_0$)	A_1	$3m + 2m_v + 2m_d + m_0 - 1$
	A_2	$3m + m_v + m_d - 1$
	B_1	$3m + 2m_v + m_d$
	B_2	$3m + m_v + 2m_d$
	E_1	$6m + 3m_v + 3m_d + m_0 - 2$
	E_2	$6m + 3m_v + 3m_d$
$C_{\infty v}$ ($N = m_0$)	Σ^+	$m_0 - 1$
	Σ^-	0
	Π	$m_0 - 2$
	$\Delta, \Phi \ldots$	0
C_{3h} ($N = 6m + 3m_h + 2m_3 + m_0$)	A'	$3m + 2m_h + m_3 - 1$
	A''	$3m + m_h + m_3 + m_0 - 1$
	E'	$3m + 2m_h + m_3 + m_0 - 1$
	E''	$3m + m_h + m_3 - 1$

Point Group, Total Number of Atoms	Species of Vibration	Number of Vibrations
C_{4h} ($N = 8m + 4m_h + 2m_4 + m_0$)	A_g	$3m + 2m_h + m_4 - 1$
	A_u	$3m + m_h + m_4 + m_0 - 1$
	B_g	$3m + 2m_h$
	B_u	$3m + m_h$
	E_g	$3m + m_h + m_4 - 1$
	E_u	$3m + 2m_h + m_4 + m_0 - 1$
C_{6h} ($N = 12m + 6m_h + 2m_6 + m_0$)	A_g	$3m + 2m_h + m_6 - 1$
	A_u	$3m + m_h + m_6 + m_0 - 1$
	B_g	$3m + m_h$
	B_u	$3m + 2m_h$
	E_{1g}	$3m + m_h + m_6 - 1$
	E_{1u}	$3m + 2m_h + m_6 + m_0 - 1$
	E_{2g}	$3m + 2m_h$
	E_{2u}	$3m + m_h$
$D_{2d} \equiv V_d (\equiv S_{4v})$ ($N = 8m + 4m_d + 4m_2$ $+ 2m_4 + m_0$)	A_1	$3m + 2m_d + m_2 + m_4$
	A_2	$3m + m_d + 2m_2 - 1$
	B_1	$3m + m_d + m_2$
	B_2	$3m + 2m_d + 2m_2 + m_4 + m_0 - 1$
	E	$6m + 3m_d + 3m_2 + 2m_4 + m_0 - 2$

Point Group, Total Number of Atoms	Species of Vibration	Number of Vibrations
$D_{3d}(\equiv S_{6v})$ $(N = 12m + 6m_d + 6m_2$ $+ 2m_6 + m_0)$	A_{1g}	$3m + 2m_d + m_2 + m_6$
	A_{1u}	$3m + m_d + m_2$
	A_{2g}	$3m + m_d + 2m_2 - 1$
	A_{2u}	$3m + 2m_d + 2m_2 + m_6 + m_0 - 1$
	E_g	$6m + 3m_d + 3m_2 + m_6 - 1$
	E_u	$6m + 3m_d + 3m_2 + m_6 + m_0 - 1$
$D_{4d}(\equiv S_{8v})$ $(N = 16m + 8m_d$ $+ 8m_2 + 2m_8 + m_0)$	A_1	$3m + 2m_d + m_2 + m_8$
	A_2	$3m + m_d + 2m_2 - 1$
	B_1	$3m + m_d + m_2$
	B_2	$3m + 2m_d + 2m_2 + m_8 + m_0 - 1$
	E_1	$6m + 3m_d + 3m_2 + m_8 + m_0 - 1$
	E_2	$6m + 3m_d + 3m_2$
	E_3	$6m + 3m_d + 3m_2 + m_8 - 1$
D_{3h} $(N = 12m + 6m_v + 6m_h$ $+ 3m_2 + 2m_3 + m_0)$	A_1'	$3m + 2m_v + 2m_h + m_2 + m_3$
	A_1''	$3m + m_v + m_h$
	A_2'	$3m + m_v + 2m_h + m_2 - 1$
	A_2''	$3m + 2m_v + m_h + m_2 + m_3 + m_0 - 1$
	E'	$6m + 3m_v + 4m_h + m_2 + 2m_2 + m_3 + m_0 - 1$
	E''	$6m + 3m_v + 2m_h + m_2 + m_3 - 1$

Point Group, Total Number of Atoms	Species of Vibration	Number of Vibrations
	A_{1g}	$3m + 2m_v + 2m_d + 2m_h + m_2 + m_2' + m_4$
	A_{1u}	$3m + m_v + m_d + m_h$
	A_{2g}	$3m + m_v + m_d + 2m_h + m_2 + m_2' - 1$
	A_{2u}	$3m + 2m_v + 2m_d + m_h + m_2 + m_2' + m_4 + m_0 - 1$
D_{4h}	B_{1g}	$3m + 2m_v + m_d + 2m_h + m_2 + m_2 + m_2'$
	B_{1u}	$3m + m_v + 2m_d + m_h + m_2'$
$(N = 16m + 8m_v + 8m_d$	B_{2g}	$3m + 2m_v + m_d + 2m_h + m_2 + m_2'$
$+ 8m_h + 4 m_2 + 4 m_2'$	B_{2u}	$3m + m_v + 2m_d + m_h + m_2$
$+ 2m_4 + m_0)$	E_g	$6m + 3m_v + 3m_d + 2m_h + m_2 + m_2' + m_4 - 1$
	E_u	$6m + 3m_v + 3m_d + 4m_h + 2m_2 + 2m_2' + m_4 + m_0 - 1$
	A_1'	$3m + 2m_v + 2m_h + m_2 + m_5$
	A_1''	$3m + m_v + m_h$
	A_2'	$3m + m_v + 2m_h + m_2 - 1$
D_{5h}	A_2''	$3m + 2m_v + m_h + m_2 + m_5 + m_0 - 1$
	E_1'	$6m + 3m_v + 4m_h + m_2 + 2m_2 + m_5 + m_0 - 1$
$(N = 20m + 10m_v + 10m_h$	E_1''	$6m + 3m_v + 2m_h + m_2 + m_5 - 1$
$+ 5m_2 + 2m_5 + m_0)$	E_2'	$6m + 3m_v + 4m_h + 2m_2$
	E_2''	$6m + 3m_v + 2m_h + m_2$

Point Group, Total Number of Atoms	Species of Vibration	Number of Vibrations
D_{6h} ($N = 24m + 12m_v + 12m_d$ $+ 12m_h + 6m_2 + 6m_2'$ $+ 2m_6 + m_0$)	A_{1g}	$3m + 2m_v + 2m_d + 2m_h + m_2 + m_2' + m_6$
	A_{1u}	$3m + m_v + m_d + m_h$
	A_{2g}	$3m + m_v + m_d + 2m_h + m_2 + m_2' - 1$
	A_{2u}	$3m + 2m_v + 2m_d + m_h + m_2 + m_2' + m_6 + m_0 - 1$
	B_{1g}	$3m + m_v + 2m_d + m_h + m_2'$
	B_{1u}	$3m + 2m_v + m_d + 2m_h + m_2 + m_2'$
	B_{2g}	$3m + 2m_v + m_d + m_h + m_2$
	B_{2u}	$3m + m_v + 2m_d + 2m_h + m_2 + m_2'$
	E_{1g}	$6m + 3m_v + 3m_d + 2m_h + m_2 + m_2' + m_6 - 1$
	E_{1u}	$6m + 3m_v + 3m_d + 4m_h + 2m_2 + 2m_2' + m_6 + m_0 - 1$
	E_{2g}	$6m + 3m_v + 3m_d + 4m_h + 2m_2 + 2m_2'$
	E_{2u}	$6m + 3m_v + 3m_d + 2m_h + m_2 + m_2'$
$D_{\infty h}$ ($N = 2m_\infty + m_0$)	Σ_g^+	m_∞
	Σ_u^+	$m_\infty + m_0 - 1$
	Σ_g^-, Σ_u^-	0
	Π_g	$m_\infty - 1$
	Π_u	$m_\infty + m_0 - 1$
	$\Delta_g, \Delta_u, \Phi_g, \Phi_u, \ldots$	0

Point Group, Total Number of Atoms	Species of Vibration	Number of Vibrations
T $(N = 12m + 6m_2 + 4m_3 + m_0)$	A	$3m + m_2 + m_3$
	E	$3m + m_2 + m_3$
	T	$9m + 5m_2 + 3m_3 + m_0 - 2$
T_d $(N = 24m + 12m_d + 6m_2 + 4m_3 + m_0)$	A_1	$3m + 2m_d + m_2 + m_3$
	A_2	$3m + m_d$
	E	$6m + 3m_d + m_2 + m_3$
	T_1	$9m + 4m_d + 2m_2 + m_3 - 1$
	T_2	$9m + 5m_d + 3m_2 + 2m_3 + m_0 - 1$
O_h $(N = 48m + 24m_h + 24m_d$ $+ 12m_2 + 8m_3$ $+ 6m_4 + m_0)$	A_{1g}	$3m + 2m_h + 2m_d + m_2 + m_3 + m_4$
	A_{1u}	$3m + m_h + m_d$
	A_{2g}	$3m + 2m_h + m_d + m_2$
	A_{2u}	$3m + m_h + 2m_d + m_2 + m_3$
	E_g	$6m + 4m_h + 3m_d + 2m_2 + m_3 + m_d$
	E_u	$6m + 2m_h + 3m_d + m_2 + m_3$
	T_{1g}	$9m + 4m_h + 4m_d + 2m_2 + m_3 + m_4 - 1$
	T_{1u}	$9m + 5m_h + 5m_d + 3m_2 + 2m_3 + 2m_4 + m_0 - 1$
	T_{2g}	$9m + 4m_h + 5m_d + 2m_2 + 2m_3 + m_4$
	T_{2u}	$9m + 5m_h + 4m_d + 2m_2 + m_3 + m_4$

Appendix 3

The Direct Sums of Excited States and Combination States of Degenerate Vibrations

Table A3.1 Symmetry Species of the Higher Vibrational Levels of Degenerate Vibrations

The numbers in front of some symbols (for example, $2E'$) indicate how many sublevels of that particular species occur if this number is greater than 1.

Point Group	Vibrational Level	Resulting States	Vibrational Level	Resulting States
$D_{3h}[C_{3v}, D_3, C_{3h}, C_3]^a$	$(e')^2$	$A_1' + E'$	$(e'')^2$	$A_1' + E'$
	$(e')^3$	$A_1' + A_2' + E'$	$(e'')^3$	$A_1'' + A_2'' + E''$
	$(e')^4$	$A_1' + 2E'$	$(e'')^4$	$A_1' + 2E'$
	$(e')^5$	$A_1' + A_2' + 2E'$	$(e'')^5$	$A_1'' + A_2'' + 2E''$
	$(e')^6$	$2A_1' + A_2' + 2E'$	$(e'')^6$	$2A_1' + A_2' + 2E'$
$D_{4h}[C_{4v}, D_4, D_{2d} \equiv V_d, C_{4h}, C_4, S_4]^b$	$(e_g)^2$	$A_{1g} + B_{1g} + B_{2g}$	$(e_u)^2$	$A_{1g} + B_{1g} + B_{2g}$
	$(e_g)^3$	$2E_g$	$(e_u)^3$	$2E_u$
	$(e_g)^4$	$2A_{1g} + A_{2g} + B_{1g} + B_{2g}$	$(e_u)^4$	$2A_{1g} + A_{2g} + B_{1g} + B_{2g}$
	$(e_g)^5$	$3E_g$	$(e_u)^5$	$3E_u$
	$(e_g)^6$	$2A_{1g} + A_{2g} + 2B_{1g} + 2B_{2g}$	$(e_u)^6$	$2A_{1g} + A_{2g} + 2B_{1g} + 2B_{2g}$
$D_{5h}[C_{5v}, D_5, C_{5h}, C_5]^c$	$(e_1)^2$	$A_1' + E_2'$	$(e_1')^2$	$A_1' + E_2'$
	$(e_1)^3$	$E_1' + E_2'$	$(e_1')^3$	$E_1'' + E_2''$
	$(e_1)^4$	$A_1' + E_1' + E_2'$	$(e_1')^4$	$A_1' + E_1' + E_2'$
	$(e_2)^2$	$A_1' + E_2'$	$(e_2')^2$	$A_1' + E_1'$
	$(e_2)^3$	$E_1' + E_2'$	$(e_2')^3$	$E_1'' + E_2''$
	$(e_2)^4$	$A_1' + E_1' + E_2'$	$(e_2')^4$	$A_1' + E_1' + E_2'$
$C_{6v}, D_6[D_{6h}, D_{3d}, C_{6h}, C_6, S_6]^d$	$(e_1)^2$	$A_1 + E_2$	$(e_2)^2$	$A_1 + E_2$
	$(e_1)^3$	$B_1 + B_2 + E_1$	$(e_2)^3$	$A_1 + A_2 + E_2$
	$(e_1)^4$	$A_1 + 2E_2$	$(e_2)^4$	$A_1 + 2E_2$
	$(e_1)^5$	$B_1 + B_2 + 2E_1$	$(e_2)^5$	$A_1 + A_2 + 2E_2$
	$(e_1)^6$	$2A_1 + A_2 + 2E_2$	$(e_2)^6$	$2A_1 + A_2 + 2E_2$

Group	Config	Reduction	Config	Reduction
$D_{4d},\,C_{8v},\,D_8$	$(e_1)^2$	$A_1 + E_2$	$(e_2)^2$	$A_1 + B_1 + B_2$
	$(e_1)^3$	$E_1 + E_3$	$(e_2)^3$	$2E_2$
	$(e_1)^4$	$A_1 + B_1 + B_2 + E_2$	$(e_2)^4$	$2A_1 + A_2 + B_1 + B_2$
	$(e_3)^2$	$A_1 + E_2$	$(e_3)^4$	$A_1 + B_1 + B_2 + E_2$
	$(e_3)^3$	$E_1 + E_3$		
$D_{\infty h}[C_{\infty v}]^{[e]}$	$(\pi_g)^2$	$\Sigma_g^+ + \Delta_g$	$(\pi_u)^2$	$\Sigma_g^+ + \Delta_g$
	$(\pi_g)^3$	$\Pi_g + \Phi_g$	$(\pi_u)^3$	$\Pi_u + \Phi_u$
	$(\pi_g)^4$	$\Sigma_g^+ + \Delta_g + \Gamma_g$	$(\pi_u)^4$	$\Sigma_g^+ + \Delta_g + \Gamma_g$
	$(\pi_g)^5$	$\Pi_g + \Phi_g + H_g$	$(\pi_u)^5$	$\Pi_u + \Phi_u + H_u$
	$(\pi_g)^6$	$\Sigma_g^+ + \Delta_g + \Gamma_g + I_g$	$(\pi_u)^6$	$\Sigma_g^+ + \Delta_g + \Gamma_g + I_g$
$T_d,\,O[O_h,\,T]^{[f]}$	$(e)^2$	$A_1 + E$	$(e)^5$	$A_1 + A_2 + 2E$
	$(e)^3$	$A_1 + A_2 + E$	$(e)^6$	$2A_1 + A_2 + 2E$
	$(e)^4$	$A_1 + 2E$	$(e)^7$	$A_1 + A_2 + 3E$
	$(f_1)^2$	$A_1 + E + T_2$	$(f_2)^2$	$A_1 + E + T_2$
	$(f_1)^3$	$A_2 + 2T_1 + T_2$	$(f_2)^3$	$A_1 + T_1 + 2T_2$
	$(f_1)^4$	$2A_1 + 2E + T_1 + 2T_2$	$(f_2)^4$	$2A_1 + 2E + T_1 + 2T_2$
	$(f_1)^5$	$A_2 + E + 4T_1 + 2T_2$	$(f_2)^5$	$A_1 + E + 2T_1 + 4T_2$
	$(f_1)^6$	$3A_1 + A_2 + 3E + 6T_1 + 4T_2$	$(f_2)^6$	$3A_1 + A_2 + 3E + 2T_1 + 4T_2$
	$(f_1)^7$	$2A_2 + 2E + 6T_1 + 4T_2$	$(f_2)^7$	$2A_1 + 2E + 4T_1 + 6T_2$

[a] For C_{3v} and D_3 the subscripts, and for C_{3h} the subscripts, and for C_3 both primes and subscripts should be omitted.

[b] For C_{4v}, D_4, and $D_{2d} \equiv V_d$ the subscripts g and u, for C_{4h} the subscripts 1 and 2, and for C_4 all subscripts should be omitted.

[c] For C_{5v} and D_5 the primes, for C_{5h} the subscripts, and for C_5 both should be omitted.

[d] For D_{6h}, D_{3d}, and S_6 the g, u rule must be taken into account; for C_{6h}, C_6, and S_6 the subscripts A and B should be omitted; for D_{2d}, B should be put equal to A, and the subscripts of E should be omitted.

[e] For $C_{\infty v}$ the subscripts should be dropped. The higher levels of π_g and π_u only are given since π vibrations are the only vibrations that occur.

[f] For O_h the g, u rule applies; for T the subscripts should be omitted.

Table A3.2 Symmetry Species of Those Levels in Which Two Different Degenerate Vibrations are Singly Excited

Point Group	Vibrational Configuration	Resulting States	Vibrational Configuration	Resulting States
$D_{3h}[C_{3v}, D_3, C_{3h}]^a$	$e' \cdot e'$	$A_1' + A_2' + E'$	$e'' \cdot e''$	$A_1' + A_2' + E'$
	$e' \cdot e''$	$A_1'' + A_2'' + E''$		
$D_{4h}[C_{4v}, D_4, D_{2d} \equiv V_d, C_{4h}, C_4, S_4]^b$	$e_g \cdot e_g$	$A_{1g} + A_{2g} + B_{1g} + B_{2g}$	$e_u \cdot e_u$	$A_{1g} + A_{2g} + B_{1g} + B_{2g}$
	$e_g \cdot e_u$	$A_{1u} + A_{2u} + B_{1u} + B_{2u}$		
$D_{5h}[C_{5v}, D_5, C_{5h}, C_5]^c$	$e_1' \cdot e_1'$	$A_1' + A_2' + E_2'$	$e_1'' \cdot e_2'$	$E_1'' + E_2''$
	$e_1' \cdot e_1''$	$A_1'' + A_2'' + E_2''$	$e_1'' \cdot e_2''$	$E_1' + E_2'$
	$e_1' \cdot e_2'$	$E_1' + E_2'$	$e_2' \cdot e_2'$	$A_1' + A_2' + E_1'$
	$e_1' \cdot e_2''$	$E_1'' + E_2''$	$e_2' \cdot e_2''$	$A_1'' + A_2'' + E_1''$
	$e_1'' \cdot e_1''$	$A_1' + A_2' + E_2'$	$e_2'' \cdot e_2''$	$A_1' + A_2' + E_1'$
$C_{6v}, D_6[D_{6h}, D_{3d}, C_{6h}, C_6, S_6]^d$	$e_1 \cdot e_1$	$A_1 + A_2 + E_2$	$e_1 \cdot e_2$	$B_1 + B_2 + E_1$
	$e_2 \cdot e_2$	$A_1 + A_2 + E_2$		
D_{4d}, C_{8v}, D_8	$e_1 \cdot e_1$	$A_1 + A_2 + E_2$	$e_2 \cdot e_2$	$A_1 + A_2 + B_1 + B_2$
	$e_1 \cdot e_2$	$E_1 + E_3$	$e_2 \cdot e_3$	$E_1 + E_3$
	$e_1 \cdot e_3$	$B_1 + B_2 + E_2$	$e_3 \cdot e_3$	$A_1 + A_2 + E_2$

Point Group	Vibrational Configuration	Resulting States	Vibrational Configuration	Resulting States
$C_{\infty v}[D_{\infty h}]^e$	$\pi \cdot \pi$	$\Sigma^+ + \Sigma^- + \Delta$	$\delta \cdot \delta$	$\Sigma^+ + \Sigma^- + \Gamma$
	$\pi \cdot \delta$	$\Pi + \Phi$	$\delta \cdot \varphi$	$\Pi + H$
	$\pi \cdot \varphi$	$\Delta + \Gamma$	$\varphi \cdot \varphi$	$\Sigma^+ + \Sigma^- + I$
$T_d, O[O_h, T]^f$	$e \cdot e$	$A_1 + A_2 + E$	$t_1 \cdot t_1$	$A_1 + E + T_1 + T_2$
	$e \cdot t_1$	$T_1 + T_2$	$t_1 \cdot t_2$	$A_2 + E + T_1 + T_2$
	$e \cdot t_2$	$T_1 + T_2$	$t_2 \cdot t_2$	$A_1 + E + T_1 + T_2$

$a-f$ See footnotes to Table A3.1.

Index